Maths Made Easy

Maths Made Easy

Dexter J. Booth

School of Computing and Mathematics
The University of Huddersfield
UK

CHAPMAN & HALL

London · Glasgow · Weinheim · New York · Tokyo · Melbourne · Madras

Published by Chapman & Hall, 2-6 Boundary Row, London SE1 8HN, UK

Chapman & Hall, 2-6 Boundary Row, London SE1 8HN, UK

Blackie Academic & Professional, Wester Cleddens Road, Bishopbriggs, Glasgow G64 2NZ, UK

Chapman & Hall GmbH, Pappelallee 3, 69469 Weinheim, Germany

Chapman & Hall USA., 115 Fifth Avenue, New York, NY 10003, USA

Chapman & Hall Japan, ITP-Japan, Kyowa Building, 3F, 2-2-1 Hirakawacho, Chiyoda-ku, Tokyo 102, Japan

Chapman & Hall Australia, 102 Dodds Street, South Melbourne, Victoria 3205, Australia

Chapman & Hall India, R. Seshadri, 32 Second Main Road, CIT East, Madras 600 035, India

First edition 1995
Reprinted 1996

© 1995 Dexter J. Booth

Printed in Great Britain at the Alden Press, Oxford

ISBN 0 412 71870 7

∞ Printed on permanent acid-free text paper, manufactured in accordance with ANSI/NISO Z39.48-1992 and ANSI/NISO Z39.48-1984 (Permanence of Paper)

Contents

How to use this book

This book is designed to be used either as a course text or as a self-study text. It is highly structured, being laid out in three parts:

Arithmetic
Algebra
Trigonometry

Each part is comprised of a number of modules where each module consists of a collection of units. Each unit follows the same pattern of exposition:

test questions

followed by:

text material

interspersed with:

worked examples and exercises.

Within a given unit, the test questions, the worked examples and the exercises are all similar in form.

The purpose of the test at the beginning of the unit is to enable you to check your knowledge and ability before reading the text. As you read the text you are expected to read through the worked examples making sure that you fully understand them; the worked examples are as valuable and important as the unit material they exemplify. Following each set of worked examples is a collection of exercises which you are expected to attempt so as to help you to make sure that you have absorbed the material of the text. Having completed a unit you are then recommended to try the test questions again. By doing this you will be able to measure the improvements in both your understanding and your manipulative ability, thereby increasing your self-confidence and enabling you to progress. Before you move on to the next unit make sure that you are completely satisfied with what you have done in the previous unit as the material is developmental unit by unit. If you find that you have difficulty remembering some of the material do not be over-concerned. What is important is to understand; to see each new idea as a logical consequence of earlier ideas, your memory will develop as you progress. Also, if you have difficulty in understanding some of the material then put the book aside for short time. You will be surprised how often the light dawns when you are away from the problem and least expecting illumination.

Having completed all the units in a module try the further exercises at the end of the module. These are similar to the test questions, examples and exercises that you have done in each unit of the module and their purpose is to test your longer-term memory. With long-term memory retention comes the confidence to tackle new areas of knowledge. Again, master one module before moving on to the next. You will soon find your memory of past work improving and your confidence growing.

Dexter J Booth
The University of Huddersfield, 1995

Part One

Arithmetic

The distinction between the idea of quantity and the idea that numbers can be used to describe quantities lies at the very heart of the intellectual development of rational ideas. The Tsimshian Indians of the Pacific North West coast understood quantity but they did not understand number. For every specific quantity they possessed seven different words. One word was used to describe the quantity of animals or flat objects, another word was used to describe the quantity of time or round objects, a third word was used to describe the quantity of trees or long objects, a fourth word described the quantity of canoes, a fifth humans, a sixth was used for measurements and a seventh was used to describe the quantity of objects that did not fit into the other six categories. Their culture was unable to generalize the concept of quantity into the concept of number; they were unable to comprehend that two animals and two canoes were both instances of the general idea of the number **two**. It is only when this generalization is understood that we are able to develop the number system independently and distinctly from the idea of quantities of physical objects.

Today we recognize the existence of number in the abstract – abstract in the sense that numbers are considered to exist independently of the numerals we use to identify them; we cannot possibly count the number of stars in our own galaxy but we do know that if we could count them there would be a number available to describe the quantity.

In this part the real numbers are developed alongside the arithmetic operations that act between them. The aim is to develop the familiar (or unfamiliar!) school arithmetic from a more mature point of view borne of past experiences. The approach is to develop the real numbers by invention inspired from an identified need. At the same time manipulative techniques are explained and exemplified. By diligent practice of the exercises you should attain not only a manipulative skill in handling arithmetic expressions but also a level of understanding that reduces conscious memory work to a minimum.

Module 1

The rational number system

OBJECTIVES

When you have completed this module you will be able to:

■ Define the structure of the rational number system

■ Combine rational numbers under the various arithmetic operations

There are three units in this module:

Unit 1: Whole numbers
Unit 2: Fractions
Unit 3: The arithmetic of rational numbers

Unit 1 Whole numbers

Try the following test:

1 Find the value of each of the following:
   (a) 32 + 307   (b) 82 + 28
   (c) 436 – 345   (d) 38 + 241 – 58 – 293 + 4

2 Find the value of each of the following:
   (a) – 28 + 75   (b) – 82 – 28

In the beginning ...

Can you remember when, as a child, you first learnt how to count? Here is a collection of marbles, let's count them. One, two, three, four, five. There are five marbles all together. The process is so simple that no-one can remember a time when he or she could not count. But have you ever listened to children learning numbers? They may be able to get up to ten and they may have heard about the number twenty but the numbers in between and beyond are something else. Tenty-five, threety-six and onety-one. They can be very humorous when you hear them but they do demonstrate a remarkable condition of the human mind. To learn how to count it is first necessary to know that you can count; it is necessary to know that there are such things as numbers. For a young child to grasp this is really quite remarkable. A young child knows that there are such things as apples because they taste nice. There are such things as walls because you have to be careful not to walk into them. But numbers? Have you ever seen a number? You will have seen a numeral but you cannot see, hear, smell, touch or feel a number. Numbers are inventions of the human imagination and their existence is in the abstract – in the mind. The fact that we have symbols to represent these numbers does not make them into concrete entities – only the symbols are concrete, the numerals 1, 2, 3 or the utterances **one, two, three**. Linking numbers to their numerals is a case of rote learning – a conscious process. Realizing that there are such things as numbers is a subconscious step – a process of the inner workings of the brain, achievable even by a young child. Isn't it a shame that in later years so many of us forget what, as a child, we instinctively understood.

Can you remember when you first learnt to add? Here are two red balls – one, two. Here are three blue balls – one, two, three. If we put them together we have five balls in total – one, two three, four, five. After this process has been repeated a number of times you begin to get the idea and in no time at all you can count. Simple you may think, but wait a minute – what exactly are you doing when you count? Subconsciously you are linking the intellectual idea that numbers exist in the abstract with the concrete numerals that are used to represent them and then, consciously, you are linking those numerals to real objects – in this case red and blue balls.

In the world of conscious experience we tend to merge the two ideas of number and numeral together with the result that when we meet different types of number we become so preoccupied with different types of numeral we lose the essence of what is happening and with that loss goes our ability properly to understand what it is that we are doing. By concentrating on the manipulation of different types of numeral we may become proficient at arithmetic but we lack the understanding that permits us to extend those arithmetic ideas into the realm of symbolic algebra. Conversely, we can spend a deal of time trying to get to grips with understanding mathematical structures thereby leaving insufficient time to learn how to manipulate these structures. Such has been the state of mathematical education during the latter half of this century. From an inherited belief that arithmetic manipulation was fundamental and all-important there was a cultural pendulum swing that demanded comprehension at the cost of manipulative skills. The outcome is a reactive belief that mathematics is hard to understand, virtually impossible to manipulate without inbuilt skills and irrelevant to the day-to-day culture.

This book seeks to redress the balance by recognizing that manipulative skills and understanding are not only necessary but interdependent. Only by practising manipulative skills on the conscious level can an understanding be developed at the subconscious level via concepts placed within a context. Constructing such a context is, of course, highly self-referencing being based upon experience which is only acquired by practising manipulative skills which requires an understanding of what exactly it is that you are doing.

Interrupting a self-referencing cycle in order to enter it successfully is not an easy task to achieve. To avoid many of the problems and pitfalls associated with defining a starting point we shall adopt an imaginary scenario in which the number system is developed from the beginning spurred on by a pragmatic requirement of need. From need follows invention and from invention follows manipulation thereby leading to understanding.

Integers, addition and subtraction

From the beginning the ancestors of *Homo sapiens* must have appreciated the quality of quantity without the ability to quantify quantity. The difference between a large pile of animal skins and a small pile would have been obvious as would the difference between two wild boar steaks as opposed to three. During the evolution of the early hominids it would soon have been realized that there was a need to quantify quantity, if only to compare the day's efforts as a hunter-gatherer or to share out the harvest as an early farmer. Whatever the cause, it is quite clear there was a need to quantify; the natural numbers came into being, ug, ug-ug, ug-ug-ug. As time progressed a need was recognized to record quantities in more permanent form – this leads naturally to the invention of the written numerals I, II, III, if only as tally scratchings on a stone wall.

The natural numbers

The natural numbers, or whole numbers, are represented by the numerals:

$$0, 1, 2, 3, 4, 5, 6, 7, 8, 9, 10, 11, \ldots$$

Their properties are quite simple:

■ *The natural numbers are ordered in that each number in the sequence is one whole number greater than the previous one.*

This order can be illustrated by representing the numbers as points on a line where each point or number is progressively annotated with the appropriate numeral. We say that the natural numbers are **plotted** as points on the line:

The line is drawn with an arrowhead pointing away from zero to indicate the direction of increasing numbers which are traditionally taken to increase from left to right; numbers to the right are greater than numbers to the left. Notice that the labelled points are equally spaced to indicate that each number is just one whole number greater than its predecessor. From this diagram we can see that:

two is **greater** than one because the number two is to the right of the number one. Also, three is greater than two and so on. We have a symbolism for this:

$2 > 1$ and $3 > 2$ where the symbol $>$ stands for **is greater than**. Conversely:

$1 < 2$ and $2 < 3$ where the symbol $<$ stands for **is less than**.

■ *The natural numbers are written with numerals arranged in place value order – units, tens, hundreds and so on*

For example, in the following arrangement of numerals:

32425

the location of a numeral within the arrangement dictates the number that it represents. For instance, the numeral 2 to the left of the numeral 4 represents the number 2000 whereas the numeral 2 to the right of the numeral 4 represents the number 20.

■ *The natural numbers are not finite in extent but infinite in extent – there is no largest natural number*

No matter how large a given natural number is there is always another natural number greater by at least one whole number.

Having invented the natural numbers and the numerals the first use that we have for them is to account for accumulations or additions. If I add these two balls to the three that I already have then I shall have five balls. Thus is born the process, or **operation** of **addition** between pairs of natural numbers. From the order property of the natural numbers we can see that:

0 ADD 1 EQUALS 1
1 ADD 1 EQUALS 2
2 ADD 1 EQUALS 3

This is tedious. We need some sort of abbreviation. Let the symbol + stand for ADD and let the symbol = stand for equals:

$0 + 1 = 1$
$1 + 1 = 2$
$2 + 1 = 3$

That is a lot better. However, by simply adding 1 in each case we are describing only the ordering of the natural numbers.

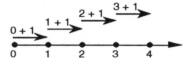

We can, however, extend this process to define the addition of any two natural numbers. For example:

Here we see that:

$2 + 3 = 5$

Indeed, by trial and error we can add any two natural numbers together to obtain another natural number. Furthermore, we can see that the process of addition has the following three properties:

(a) If we add two natural numbers together the result is another natural number

(b) The order in which we add two numbers together does not matter. We always obtain the same result. For example:

$2 + 3 = 3 + 2 = 5$

(c) The way we associate three or more numbers in an addition does not matter, we always obtain the same result. For example:

$$2 + (3 + 4) = (2 + 3) + 4$$
$$= 9$$

Notice that here we have introduced the bracket (...) which is a device used to indicate which addition is to be performed first.

We always execute the addition inside the bracket before we execute the addition outside the bracket.

Now we have the beginnings of what, in mathematics, is called a **structure**. Every mathematical structure consists of:

objects
properties assigned to those objects
operations between objects
a collection of rules that are obeyed whenever objects are operated upon.

Here, our objects are the natural numbers and the properties they possess are that they are ordered, infinite in extent and are written using numerals arranged in place value order. The operation is addition and the rules that we have so far are that:

■ *Any two natural numbers can be added together to give another natural number*

■ *The order in which two natural numbers are added does not matter.*

■ *The way in which three or more natural numbers are associated in an addition does not matter.*

Notice that in all that has been said so far we have not **defined** either the natural numbers or the operation of addition. Any definition rests upon earlier concepts and in the case of the natural numbers and addition there are no earlier concepts. Our discussion so far rests upon a consensual understanding of what it is we are talking about. From here on in we can define everything.

Having invented the operation of addition to account for accumulations it immediately becomes necessary to invent an operation that accounts for subtractions – taking away is as natural a process as adding to. So the operation of subtraction is born, the **reverse operation to addition**:

    3 TAKE AWAY 1 = 2
    2 TAKE AWAY 1 = 1
    1 TAKE AWAY 1 = 0
    0 TAKE AWAY 1 = ?

or, better still:

    3 – 1 = 2
    2 – 1 = 1
    1 – 1 = 0
    0 – 1 = ?

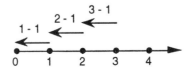

Hello – a problem! We cannot take one away from nothing. We found earlier that we could add *any* two natural numbers together to obtain another natural number but we now find that we cannot do the same with subtraction. We cannot, for example take one away from nothing; we do not have a natural number that fits the bill. At this point we must look to the structure that we have so far developed because already the fledgling is showing signs of weakness. Weakness in the sense that a restriction is imposed on the numbers combined in a subtraction whereas no restriction is imposed on the numbers that can be combined under addition; this despite that subtraction is merely the reverse process to addition. The essence of the problem is that we cannot take a larger number away from a smaller number but do we need to do this anyway? If we are only going to use the numbers to count complete physical objects then we need proceed no further; our number system is adequate. However, the structure has a restriction that is aesthetically unpleasing – besides, we all know that we need to be able to take larger numbers away from smaller numbers because we all borrow money.

Putting on our pragmatic hat we overcome the problem by **inventing** a new kind of number – a **negative whole number**. We define the number **negative-one** to be that number that is the result of taking 1 away from nothing. In numerals we define:

$$0 - 1 \text{ to be } -1$$

that is:

$$0 - 1 = -1$$

where –1 is the numeral for this example of this new type of number. Indeed, we can create a negative whole number corresponding to every natural number. For example:

$$0 - 24 \ = -24$$
$$0 - 111 = -111 \text{ and so on.}$$

Graphically, this means that we have extended the line of natural numbers to the left of zero to accommodate these negative whole numbers:

$$-4 \quad -3 \quad -2 \quad -1 \quad 0 \quad 1 \quad 2 \quad 3 \quad 4$$

Adding a negative integer is now seen to be akin to subtracting a positive integer:

$$8 + (-3) = 8 - 3$$
$$= 5$$

We can now solve the problem of taking a larger number from a smaller number. For example, to work out the value of:

$$15 - 20$$

we recognize that:

$$20 = 15 + 5$$

so when we subtract 20 from 15 we first of all subtract 15 to leave 0;

$$15 - 15 = 0$$

We still have 5 left to subtract so:

$$0 - 5 = -5$$

The final result.

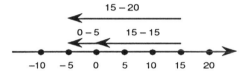

Subtracting a negative integer requires some thought. For example, what is the value of:

$$3 - (-2)?$$

If we write 3 as $5 + (-2)$ then we find that:

$$3 - (-2) = 5 + (-2) - (-2)$$
$$= 5$$
$$= 3 + 2$$

Consequently, we find that subtracting a negative integer is akin to adding a positive integer.

How is our mathematical structure looking now?

■ *We have a collection of objects, the natural numbers and the negative whole numbers. We call the entire collection the integers – the positive and negative integers.*

■ *The integers share the same properties of the natural numbers in that they are ordered, infinite in extent and written using the numerals located in place value order.*

■ *We have two operations of addition and subtraction between integers with the following property:*

Adding or subtracting two integers results in another integer.

The order in which we add two integers or associate three or more integers in an addition does not matter. However, the order in which we subtract two integers or associate three or more integers in a subtraction does matter. For example:

$$5 - 3 \neq 3 - 5$$

Notice the use of the symbol $\neq$ which stand for **does not equal**.
Also:

$$[6 - (-4)] - 2 = (6 + 4) - 2 = 10 - 2 = 8$$

whereas:

$$6 - [(-4) - 2] = 6 - (-6) = 6 + 6 = 12$$

Worked Examples

1.1 Find the value of each of the following:
  (a) $13 + 246$ (b) $19 + 91$
  (c) $245 - 154$ (d) $111 + 345 - 832 - 932 + 5$

Solution:
(a) $13 + 246$ is written as:

$$\begin{array}{r} 13 \\ + 246 \\ \hline 259 \end{array}$$

By adding numbers in columns. Alternatively, we reason that:

$13 = 10 + 3$ and
$246 = 200 + 40 + 6$ by expanding the place values. Therefore:

$$\begin{aligned} 13 + 246 &= (10 + 3) + (200 + 40 + 6) \\ &= 200 + (40 + 10) + (6 + 3) \\ &= 200 + 50 + 9 \\ &= 259 \end{aligned}$$

(b) $\begin{aligned} 19 + 91 &= (10 + 9) + (90 + 1) \\ &= (90 + 10) + (9 + 1) \\ &= 100 + 10 \\ &= 110 \end{aligned}$

(c) $\begin{aligned} 245 - 154 &= (200 + 40 + 5) - (100 + 50 + 4) \\ &= (200 - 100) + (40 - 50) + (5 - 4) \\ &= 100 + (-10) + 1 \\ &= (100 - 10) + 1 \end{aligned}$

$$= 90 + 1$$
$$= 91$$

Alternatively:

$$245 - 154 = 245 - (145 + 9)$$
$$= 245 - 145 - 9$$
$$= 100 - 9$$
$$= 91$$

(d) $111 + 345 - 832 - 932 + 5 = (111 + 345 + 5) - (832 + 932)$
$$= 461 - 1764$$
$$= 464 - 3 - 1000 - 764$$
$$= -1000 - 764 + 464 - 3$$
$$= -100 - 300 - 3$$
$$= -1303$$

1.2 Find the value of each of the following:
   (a) $-13 + 246$ (b) $-19 - 91$

Solution:
(a) $-13 + 246 = 246 - 13 = 233$

(b) $-19 - 91 = -(19 + 91) = -110$

Exercises

1.1 Find the value of each of the following:
   (a) $24 + 105$ (b) $37 + 73$
   (c) $305 - 124$ (d) $87 + 179 - 23 - 566 + 8$

1.2 Find the value of each of the following:
   (a) $-24 + 105$ (b) $-37 - 73$

Unit 2   Fractions

Try the following test:

1   Find the value of each of the following:
    (a) 2341×11 (b) 39×684

2   Find the lowest common multiple (LCM) of each of the following
    pairs of numbers:
    (a) 4 and 5   (b) 6 and 18   (c) 15 and 20

3   Find the value of each of the following:
    (a) $(-2341) \times 11$ (b) $(-39) \times (-684)$

4   Find the value of each of the following:
    (a) $273 \div 3$ (b) $3774/17$
    (c) $273 \div (-3)$ (d) $(-3774)/(-17)$

5   Find the factors and highest common factor (HCF) of 18 and 45

6   Find the prime factors of each of the following:
    (a) 630   (b) 273

Multiplication
As time progresses it becomes a universal gripe that all this repetitive addition is
becoming very tedious and boring:

$$3 + 3 + 3 + 3 + 3 = 15$$

Someone comes up with a new idea. Why not define a new operation that takes
account of the number of times a given number is repeated in a sum such as the one
above. We shall call the operation **multiplication** to indicate the multiple nature of
the addition:

$$3 \text{ MULTIPLIED BY } 5 = 15$$

If we *remember* that 3 MULTIPLIED BY 5 is equal to 15 then we have invented a
shortcut to repetitive addition. And that is where the **multiplication tables** come in
and why you had to spend so many hours trying memorize them – to enable you to
add more efficiently. Just as with addition and subtraction the multiplication
operation has a symbol to represent it $\times$, so that we can write:

$$3 \times 5 = 15$$

We call 15 the **product** of 3 and 5. Again, just as with addition and subtraction, multiplying two integers produces another integer. Furthermore, the order in which two integers are multiplied or the order in which three or more integers are associated in a multiplication does not matter. For example:

$$3 \times 5 = 5 \times 3$$

that is:

$$3 + 3 + 3 + 3 + 3 = 5 + 5 + 5$$

(three rows of five dots total to the same number of dots as five columns of three dots)

$$
\begin{matrix}
\bullet & \bullet & \bullet & \bullet & \bullet \\
\bullet & \bullet & \bullet & \bullet & \bullet \\
\bullet & \bullet & \bullet & \bullet & \bullet
\end{matrix}
$$

and

$$(4 \times 6) \times 7 = 4 \times (6 \times 7)$$

All progresses well until someone points out that we are going to have a problem with:

$$3 \times (-5)$$

A little bit of thought shows that we can write:

$$3 \times (-5) = (-5) \times 3 = (-5) + (-5) + (-5) = -15$$

In particular, for example:

$$4 \times (-1) = (-1) \times 4 = -4$$

>*Multiplying a number by –1 changes the sign of the number.*

Following this statement through logically means that, for example:

$$(-1) \times (-4) = 4$$

>*The product of two negative integers is a positiver integer.*

In particular:

$$(-1) \times (-1) = 1$$

When this definition is first met it can cause quite a number of conceptual difficulties.

It does not seem to make sense – how can we multiply a negative number by a negative number and get a positive number? Difficulties such as this one arise because the counting aspect of numbers has been over-emphasized. Whilst it is agreed that the existence of the natural numbers enables us to count real objects the extension of the number system to include the negative whole numbers is a pragmatic one that is devised to make the mathematical structure more robust. The fact that we can use the negative whole numbers to describe debts and freezing temperatures is a bonus. Indeed, there was a time when these negative whole numbers were referred to as **impossible numbers**. We must at all times remember that the number system exists as a mental construct independent of any practical use it may have. Consequently, when we define:

$$(-1) \times (-1) = 1$$

we do so in order to develop a consistent, abstract number system.

We can now multiply two negative numbers quite happily:

$$
\begin{aligned}
(-3) \times (-5) &= [(-1) \times 3] \times [(-1) \times 5] \\
&= (-1) \times 3 \times (-1) \times 5 &&\text{removing the square brackets} \\
&= (-1) \times (-1) \times 3 \times 5 &&\text{the order of multiplication does} \\
&= 1 \times 3 \times 5 &&\text{not matter} \\
&= 15
\end{aligned}
$$

Having overcome the problems of multiplying by negative numbers we still have one more little problem to clear up. What is the value of:

$$5 \times 0?$$

Again:

$$
\begin{aligned}
5 \times 0 &= 0 \times 5 \\
&= 0 + 0 + 0 + 0 + 0 \\
&= 0 &&\text{any number multiplied by 0 is 0. Even } 0 \times 0 = 0
\end{aligned}
$$

Factors and prime numbers
If a natural number can be expressed as a product of two or more natural numbers the natural numbers in the product are called **factors**. For example:

$$12 = 3 \times 4$$

so that 3 and 4 are factors of 12. Notice that they are not the only factors of 12 because:

$$12 = 6 \times 2$$

and

$$12 = 12 \times 1$$

so even 12 and 1 are factors of 12.

If a natural number only has two factors, namely itself and unity, then that natural number is called a **prime number**. For example:

2, 3, 5, 7, 11, 13, 17

are the first seven prime numbers because their only factors are themselves and unity. Notice that:

1

is not a prime number because it only has one factor – itself.

Worked Examples

1.3 Find the value of each of the following:
    (a) 672×11 (b) 29×387

Solution:
(a) $672 \times 11 = 672 \times (10 + 1)$
$$= (672 \times 10) + (672 \times 1)$$
$$= 6720 + 672$$
$$= 7392$$

(b) $29 \times 387 = (30 - 1) \times 387$
$$= (30 \times 387) - (1 \times 387)$$
$$= 11610 - 387$$
$$= 11223$$

1.4 Find the lowest common multiple (LCM) of each of the following pairs of numbers:
    (a) 3 and 5   (b) 8 and 24   (c) 12 and 18

The lowest common multiple (LCM) of two given numbers is the smallest number for which both given numbers are factors.

Solution:
(a) The smallest number for which 3 and 5 are both factors is 15. Notice:

$3 \times 5 = 15$

(b) $24 = 8 \times 3$ so that 24 is the LCM of 8 and 24

(c) $12 = 2 \times 2 \times 3$ the prime factorization of 12

$18 = 2 \times \quad 3 \times 3$ the prime factorization of 18

$36 = 2 \times 2 \times 3 \times 3$ the lowest common multiple (LCM) of 12 and 18

Here the two factors 2 of 12 and the two factors 3 of 18 must both appear in the common multiple.

1.5 Find the value of each of the following:
     (a) $(-672) \times 11$ (b) $(-29) \times (-387)$

Solution:

(a) $(-672) \times 11 = (-1) \times 672 \times 11 = -7392$

(b) $(-29) \times (-387) = (-1) \times 29 \times (-1) \times 387 = (-1) \times (-1) \times 29 \times 387 = 11223$

Exercises

1.3 Find the value of each of the following:
     (a) 1339×9 (b) 51×403

1.4 Find the lowest common multiple (LCM) of each of the following pairs of numbers:
     (a) 7 and 9   (b) 5 and 15   (c) 16 and 24

1.5 Find the value of each of the following:
     (a) $(-1339) \times 9$ (b) $(-51) \times (-403)$

Division

Just as we defined subtraction as the reverse process to addition so we define division as the reverse process to multiplication, namely, repetitive subtraction. Given that when we add five 2's together we obtain the number 10 we can say that 10 is made up of five 2's. That is, if we take five 2's away from 10 we are left with nothing. The process of division, counting the number of times one integer can be taken from another integer, is also referred to as the **quotient** from the Middle English word **quotiens** meaning **how many times**.

Symbolically:

$$2 + 2 + 2 + 2 + 2 = 10 \text{ that is } 2 \times 5 = 10$$

so

$$10 - 2 - 2 - 2 - 2 - 2 = 0$$

This fact we write as:

$$10 \div 2 = 5$$

where the symbol ÷ stands for **division** – repetitive subtraction. A more convenient symbolism for division is to write it as a **ratio** of integers, for example:

$$\frac{10}{2} = 5$$

The reason we alternatively write the division, or **quotient** of two numbers as a ratio is because it makes the manipulation of such quotients so much more intuitive as we shall observe as we progress. In a text such as this, however, the format of the ratio is often written as, for example:

$$10/2 = 5$$

where the forward slash (/) represents the operation of division. This alternative form was originally as much for the convenience of the typesetter until the advent of the computer where the forward slash is now used to represent division. In what follows we shall use all three notations, selecting that which is more appropriate at the time. Just as we found problems with the operation of subtraction so we have problems with the operation of division. We find that the integers are not closed under the operation of division. For example, what does

$$3 \div 6$$

mean? We cannot take 6 from 3 to leave 0 even once. Again, just as we found with subtraction, we have come across a well defined operation between two numbers, the result of which raises a restriction on the numbers to which it can be applied. There is no integer that is equal to:

$3 \div 6$ or, alternatively, $\dfrac{3}{6}$ (or 3/6 in the other notation)

Whenever we define a new operation it will naturally be restricted by the rules of applicability imposed upon it. However, to restrict the objects on which it operates in this somewhat arbitrary way is unsatisfactory. We overcame a similar problem with the operation of subtraction by extending the number system and this is what we do now. We extend the number system to include almost all ratios of integers as bona fide numbers. For example, we accept that:

$$1/2, \ -3/6, \ (-5)/(-3), \ 18/(-7)$$

are all bona fide numbers and as such can be plotted on our line of numbers.

The only ratios that we do not accept are those where the number on the bottom is zero. For example:

$$5 \div 0$$

is not defined because no matter how many times 0 is subtracted from 5 it is not possible to end up with 0.

We have **invented** a new kind of number – a **ratio of integers** where the integer on the top is called the **numerator** and the integer on the bottom is called the **denominator**. We call all such ratios of integers **rational** numbers where the word rational springs from the word **ratio** and not from any other use of the word rational. An added bonus to our pragmatic invention of the rational numbers is that we can use them in our counting process. If a whole cake is cut into four equal parts then each part is:

$$1/4$$

one fourth (or one quarter) of the whole. Hence the alternative name for ratios of integers as **fractions** because they can be used to represent fractional parts of a whole.

Notice that the integers can be considered as rational numbers themselves. For instance:

$$35 = \frac{35}{1} \text{ or even } \frac{70}{2}$$

Worked Examples

1.6 Find the value of each of the following:
(a) $832 \div 4$ (b) $(-34187)/(-17)$
(c) $832 \div (-4)$ (d) $34187/17$

Solution:
(a) $832 \div 4 = (800 + 30 + 2) \div 4$
$\qquad = (800 \div 4) + (32 \div 4)$
$\qquad = 200 + 8$
$\qquad = 208$

(b) $(-34187)/(-17) = [(-1) \times 34187]/[(-1) \times 17] = 34187/17 = 2011$

(c) $832 \div (-4) = 832 \div [(-1) \times 4] = (-1) \times (832 \div 4) = -208$

Notice that because $1 = (-1) \times (-1)$ so $1 \div (-1) = (-1)$

(d) $34187/17 = (34000 + 170 + 17) \div 17$
$\qquad = 2000 + 10 + 1$
$\qquad = 2011$

Alternatively we can perform the long division:

$$\begin{array}{r} 2011 \\ 17\overline{\smash{\big)}34187} \\ 34000 \\ \hline 187 \\ 170 \\ \hline 17 \\ 17 \\ \hline 0 \end{array}$$

Subtract

$2000 \times 17 = 34000$

$10 \times 17 = 170$

$1 \times 17 = 17$

Consequently:

$34187 - (2000 + 10 + 1) \times 17 = 0$ (division is repetitive subtraction)

That is:

$34187 = 2011 \times 17$

Hence

$$\frac{34187}{17} = \frac{(2011 \times 17)}{17} = 2011$$

Make sure that you understand that this long division is indeed repetitive subtraction. You will need to understand this when we apply this method later to perform algebraic long division.

1.7 Find the factors and highest common factor (HCF) of 12 and 30

Solution:
A factor of a given natural number is a natural number that divides the given number a natural number of times.

The factors of 12 are:

1, 2, 3, 4, 6 and 12 because they all divide into 12 a whole number of times.

the factors of 30 are:

1, 2, 3, 5, 6, 10, 15 and 30

The highest common factor (HCF) of two numbers is the largest factor they have in common. In this case it is 6.

1.8 Find the prime factors of each of the following:
(a) 210   (b) 143

Solution:

A prime number is a natural number other than 1 which has no factors other than 1 and itself. For example the first seven prime numbers are:

2, 3, 5, 7, 11, 13, 17

(a) Perform the division of 210 by successive prime numbers:

```
2 | 210
3 | 105
5 |  35
7 |   7
  |   1
```

The prime factors of 210 are then 2, 3, 5 and 7.

Notice: The successive multiplication:

$$2 \times 3 \times 5 \times 7 = 210$$

This is called the **prime factorization** of 210.

(b) $143 = 11 \times 13$ and 11 and 13 are both prime numbers

---

Exercises

1.6 Find the value of each of the following:
  (a) $1025 \div 5$ (b) $8502/13$
  (c) $1025 \div (-5)$ (d) $(-8502)/(-13)$

1.7 Find the factors and highest common factor (HCF) of 18 and 45

1.8 Find the prime factors of each of the following:
  (a) 378   (b) 1485

Unit 3 The arithmetic of rational numbers

Try the following test:

1 Which pair in each of the following are equivalent fractions:
  (a) 3/4, 9/16   (b) 2/3, 16/24
  (c) −7/9, −35/45   (d) 8/11, 32/44

2 Find the value of each of the following:
  (a) $1/2 \times 1/7$ (b) $-3/8 \div 6/9$

3 Find the value of each of the following:
  (a) 1/2 + 1/7   (b) 3/8 + 5/16
  (c) 7/11 − 8/22   (d) −3/15 + 7/25

4 Find the reciprocals of each of the following:
  (a) 4/5   (b) −9/2
  (c) 2/(−7)   (d) −11

5 Find the value of:
  (a) 2/7 ÷ 1/5   (b) (−5/16)/(3/8)

6 Find the value of each of the following by hand, expressing the result as a rational number (check your results using a hand calculator):
  (a) $7 + 3 \times 2 \div 4 - 3 \div 5$ (b) $(7 + 3) \times 2 + (4 - 3) \div 5$
  (c) $7 + 3 \times 2 \div 4 - (3 \div 5)$

7 Convert the following fractions to percentages:
  (a) 3/8   (b) −2/5
  (c) 2/9   (d) 11/8

8 Convert each of the following percentages to a fraction in its lowest terms:
  (a) 20%   (b) $18\frac{3}{4}\%$
  (c) −15%   (d) 175%

Equivalent fractions and fraction arithmetic
We need to look at fractions, ratios of integers, a little more closely because there are one or two problems to clear up before we can proceed.

Firstly, we need to point out that the numeral representation of a fraction is not unique. In other words, different ratios of integers can represent the same number. For example:

$$10 \div 2 = 10/2 = 5$$
$$20 \div 4 = 20/4 = 5$$
$$5 \div 1 = 5/1 = 5$$

All three fractions 10/2, 20/4 and 5/1 represent the same integer, namely 5 and for this reason we call these three fractions **equivalent fractions**. Armed with this fact we can now proceed to see how fractions fit into our earlier work.

Worked Examples

1.9 Which pair in each of the following are equivalent fractions:
  (a) 2/3, 8/12   (b) 5/7, 25/49
  (c) −7/3, −35/15   (d) 7/3, −35/15

Solution:
(a) 2/3 and 8/12 are equivalent fractions because
   $8/12 = (2 \times 4)/(3 \times 4) = 2/3$

(b) 5/7, 25/49 are not equivalent fractions because
   $25/49 = (5 \times 5)/(7 \times 7) \neq 5/7$

(c) −7/3, −35/15 are equivalent fractions because
   $-35/15 = (-1) \times (7 \times 5)/(3 \times 5) = -7/3$

(d) 7/3, −35/15 are not equivalent fractions because
   $-35/15 = -7/3 \neq 7/3$

Exercises

1.9 Which pair in each of the following are equivalent fractions:
  (a) 3/5, 9/25   (b) 2/9, 12/54
  (c) −1/8, −12/96   (d) 7/5, −35/25

Multiplication
Multiplying two fractions cannot immediately be achieved by using the principle of repetitive addition. For example, repetitive addition will not enable us to find the value of:

$$\frac{3}{4} \times \frac{2}{5}$$

To overcome this problem we rewrite it as follows:

$$\frac{3}{4} \times \frac{2}{5} = \frac{3 \times 2}{4 \times 5}$$

Now repetitive addition can be employed to give:

$$\frac{3 \times 2}{4 \times 5} = \frac{6}{20}$$

Notice that if we multiply the numerator and the denominator of a fraction by the same number we obtain an equivalent fraction. For example:

$$\frac{7 \times 4}{3 \times 4} = \frac{28}{12}$$

which is equivalent to:

$$\frac{7}{3}$$

because:

$$\frac{7 \times 4}{3 \times 4} = \frac{7}{3} \times \frac{4}{4} = \frac{7}{3} \times 1 = \frac{7}{3}$$

Worked Examples

1.10 Find the value of each of the following:
    (a) $1/2 \times 1/3$ (b) $-7/9 \times 4/3$

Solution:
(a) $1/2 \times 1/3 = (1 \times 1)/(2 \times 3) = 1/6$

(b) $-7/9 \times 4/3 = ((-7) \times 4)/(9 \times 3) = -28/27$

Exercises

1.10 Find the value of each of the following:
    (a) $1/3 \times 1/5$ (b) $-11/3 \times 5/12$

Addition and subtraction
Adding or subtracting two fractions can be achieved only if they both possess the same denominator. Consequently, to add two fractions with different denominators equivalent fractions must be found that have the same denominators. For example, to perform the addition:

$$\frac{3}{4} + \frac{2}{5}$$

it is noted that:

$$\frac{3}{4} = \frac{3}{4} + \frac{5}{5} = \frac{15}{20}$$

and

$$\frac{2}{5} = \frac{2}{5} + \frac{4}{4} = \frac{8}{20}$$

so that:

$$\frac{3}{4} + \frac{2}{5} = \frac{15}{20} + \frac{8}{20} = \frac{23}{20}$$

Finding such a common denominator can be achieved by multiplying the two denominators together. In this case, $4 \times 5 = 20$. Similarly, for subtraction:

$$\frac{3}{4} - \frac{2}{5} = \frac{15}{20} - \frac{8}{20} = \frac{7}{20}$$

Worked Examples

1.11 Find the value of each of the following:

    (a) $1/2 + 1/3$ (b) $3/4 + 5/6$

    (c) $7/9 - 4/3$ (d) $-7/12 + 5/18$

Solution:

(a) $1/2 + 1/3 = 3/6 + 2/6 = 5/6$

(b) $3/4 + 5/6 = 9/12 + 10/12 = 19/12$ **Notice:** 12 is the LCM of 4 and 6

(c) $7/9 - 4/3 = 7/9 - 12/9 = -5/9$

(d) $-7/12 + 5/18 = -21/36 + 10/36 = -11/36$

Exercises

1.11 Find the value of each of the following:

    (a) $1/3 + 1/5$ (b) $2/5 + 3/2$

    (c) $11/3 - 5/12$ (d) $-3/16 + 9/12$

Reciprocals

If the numerator and the denominator in a fraction are interchanged the number that results is called the **reciprocal** of the original number. For example, the reciprocals of:

7/6, 2, −5 and −1/3

are, respectively:

6/7, 1/2, −1/5 and −3/1 = −3

Notice that multiplying a number by its reciprocal always results in the number 1. The notion of the reciprocal is required to enable the operation of division of fractions.

Worked Examples

1.12 Find the reciprocals of each of the following:
    (a) 2/3   (b) −5/4
    (c) 6/(−15)   (d) 217

Solution:
(a) The reciprocal of 2/3 is 3/2, obtained by interchanging the numerator and the denominator of the original fraction.

(b) The reciprocal of −5/4 is 4/(−5) = −4/5

(c) The reciprocal of 6/(−15) = −6/15 is −15/6

(d) The reciprocal of 217 is 1/217

Exercises

1.12 Find the reciprocals of each of the following:
    (a) 9/7   (b) −3/11
    (c) 1/(−9)   (d) −35

Division

Dividing a number by a fraction is equivalent to multiplying the number by the reciprocal of the fraction. For example:

$$4 \div \frac{1}{2} = 8$$

there are 8 halves in 4 wholes. Notice that:

$$4 \times 2 = 8$$

so that

$$4 \div \frac{1}{2} = 4 \times \frac{2}{1} = 4 \times 2$$

Generally, this is the case. For example:

$$\frac{5}{7} \div \frac{9}{2} = \frac{5}{7} \times \frac{2}{9} = \frac{10}{63}$$

Worked Examples

1.13 Find the value of each of the following:
    (a) 3/4 + 5/6   (b) (−7/12)/(5/18)

Solution:
(a) 4/3 ÷ 5/6 = 4/5 × 6/5
          = 24/25
(b) (−7/12)/(5/18) = (−7/12) × (18/5)
                 = −126/60
                 = −21/10

Exercises

1.13 Find the value of:
    (a) 2/5 + 3/2   (b) (−13/16)/(9/12)

Precedence rules
If an arithmetic expression contains a mixture of different operations the natural question to ask is:

Which operation do we perform first?

For example, if you were asked to find the value of:

$$3 + 4 \times 5$$

you might think that there are two possible answers:

$$3 + 4 \times 5 = 7 \times 5 = 35 \qquad \text{by adding before multiplying}$$

and

$$3 + 4 \times 5 = 3 + 20 = 23 \qquad \text{by multiplying before adding}$$

Which answer is correct? Either one could be chosen to be correct but whichever one is chosen a reason must be given. By accepted convention we agree that when multiple operations are involved in this way the following precedence rules dictate the order in which operations are performed:

Brackets	**B**
Of	**O**
Division	**D**
Multiplication	**M**
Addition	**A**
Subtraction	**S**

the so-called BODMAS rules. This means that:

$$3 + 4 \times 5 = 3 + 20 = 23.$$

because we must multiply before we add. To produce the other answer we must use brackets:

$$(3 + 4) \times 5 = 7 \times 5 = 35$$

because, by the precedence rules, brackets are evaluated first. By the way, the operation **of** is an alternative form of multiplication:

$$7 \text{ of } 9 = 7 \times 9 = 63$$

Worked Examples

1.14 Find the value of each of the following by hand, expressing the result as a rational number (check your results using a hand calculator):
    (a) $8 - 4 \div 2 \times 3 - 6 + 8 \times 3$ (b) $(8 - 4) \times 3 \div (2 - 6 + 8) \times 3$
    (c) $8 - 4 \times 3 \div (2 - 6) + 8 \times 3$

Solution:
(a) $8 - 4 \div 2 \times 3 - 6 + 8 \times 3 = 8 - 2 \times 3 - 6 + 8 \times 3$
$$= 8 - 6 - 6 + 24$$
$$= 20$$

(b) $(8 - 4) \times 3 \div (2 - 6 + 8) \times 3 = 4 \times 3 \div 4 \times 3$
$$= 1$$

(c) $8 - 4 \times 3 \div (2 - 6) + 8 \times 3 = 8 - 4 \times 3 \div (-4) + 8 \times 3$
$$= 8 - 12 \div (-4) + 24$$
$$= 8 - (-3) + 24$$
$$= 35$$

Exercises

1.14 Find the value of each of the following by hand, expressing the result as a rational number (check your results using a hand calculator):

(a) $9 + 5 \times 3 + 5 + 4 - 8 \div 2$ (b) $(9 + 5) \times 3 + (5 + 4 - 8) \div 2$

(c) $9 + 5 \times 3 + 5 + (4 - 8) \div 2$

Percentages

A **percentage** is a fractional part of 100. For example:

$$\frac{25}{100}$$

is alternatively described as twenty-five percent and written as:

$$25\%$$

Any fraction can be converted to a percentage by finding the equivalent fraction whose denominator is equal to 100. For example, the fraction:

$$\frac{2}{5}$$

has the equivalent fraction:

$$\frac{2 \times 20}{5 \times 20} = \frac{40}{100}$$

which is:

$$40\%$$

This number 40 can be obtained by multiplying the original fraction by 100:

$$\frac{2}{5} = \frac{40}{100} \text{ so that } \frac{2}{5} \times 100 = 40$$

Worked Examples

1.15 Convert the following fractions to percentages:

(a) 3/4   75%   (b) –1/2   –50%

(c) 2/3   67%   (d) 6/5

Solution:

(a) $\frac{3}{4} \times 100 = \frac{300}{4} = 75$. Therefore:

$6 \times 30 = 180$

$5 \times 30 = 150$

$$\frac{3}{4} = \frac{75}{100} = 75\%$$

(b) $-1/2 \times 100 = -50$: $-1/2 = -50\%$

(c) $2/3 \times 100 = 66\frac{2}{3}$: $2/3 = 66\frac{2}{3}\%$

(d) $6/5 \times 100 = 120$: $6/5 = 120\%$

1.16 Convert each of the following percentages to a fraction in its lowest terms:
   (a) 25%   (b) $37\frac{1}{2}\%$
   (c) −18%   (d) 150%

Solution:
(a) $25\% = 25/100 = 1/4$

(b) $37\frac{1}{2}\% = 75/200 = 3/8$

(c) $-18\% = -18/100 = -9/50$

(d) $150\% = 150/100 = 3/2$

Exercises
1.15 Convert the following fractions to percentages:
   (a) 5/8   (b) −1/4
   (c) 1/6   (d) 5/4

1.16 Convert each of the following percentages to a fraction in its lowest terms:
   (a) 15%   (b) $33\frac{1}{3}\%$
   (c) −20%   (d) 125%

Module 1 Further exercises

1 Find the value of each of the following:
   (a) 193 – 137   (b) 733 + 589 – 591 – 17
   (c) 10157 – 9432 + 349   (d) 29091 – 18980 – 73 + 4903

2 Find the value of each of the following:
   (a) –379 + 93   (b) –1703 + 675

3 Find the value of each of the following:
   (a) 2704 × 37   (b) 117 × 5936

4 Find the lowest common multiple (LCM) of each of the following pairs
   of numbers:
   (a) 18 and 24   (b) 17 and 19
   (c) 25 and 15

5 Find the value of each of the following:
   (a) (–307) × 1179   (b) (–791) × (–531)

6 Find the value of each of the following:
   (a) 9747 ÷ 19   (b) 39100/23
   (c) (–256) ÷ (–16)   (d) 1024/(–32)

7 Find the factors and highest common factor (HCF) of 18 and 45

8 Find the prime factors of each of the following:
   (a) 1287   (b) 71383

9 Which pair in each of the following are equivalent fractions:
   (a) –13/5, –26/15   (b) 9/4, 153/68
   (c) 19/23, –95/(–115)   (d) 6/8, 10/12

10 Find the value of each of the following:
   (a) (–3/12) × (–4/9)   (b) (–17/11) ÷ (–34/33) (c) 28/72 × 12/7

11 Find the value of each of the following:
   (a) 12/7 + 9/11   (b) –7/13 + 11/26
   (c) 16/24 – 3/80   (d) –15/36 – 11/27

12 Find the reciprocals of each of the following:
   (a) –7   (b) (–4)/(–9)
   (c) 1   (d) –1

13 Find the value of:
    (a) $(-17/11) \div (-34/33)$ (b) $(-54/81)/(-3/243)$

14 Find the value of each of the following by hand, expressing the result as
    a rational number (check your results using a hand calculator):
    (a) $(-16) \times 11 + (-4) + 13 - 9 \div (-3)$
    (b) $(-16) \times 11 + ((-4) + 13) - 9 \div (-3)$
    (c) $(-16) \times 11 + ((-4) + 13 - 9) \div (-3)$

15 Convert the following fractions to percentages:
    (a) 3/16   (b) $-1/8$
    (c) 5/9   (d) 7/5

16 Convert each of the following percentages to a fraction in its lowest terms:
    (a) 45%   (b) $11\frac{1}{9}\%$
    (c) -60% (d) 250%

Module 2

The irrational and the real numbers

OBJECTIVES

When you have completed this module you will be able to:

■ Define the structure of the irrational and the real number systems

■ Manipulate the arithmetic of powers

There are two units in this module:

Unit 1: Integer powers
Unit 2: Rational powers

Unit 1 Integer powers

Try the following test:

·1 By considering powers of 3 demonstrate how you could, by adding powers, show that:

$$9 \times 27 = 243$$

2 Show that a natural number raised to the fourth power is also the square of that number squared.

3 Find the value of each of the following:
(a) 5^3 (b) 6^{-2}
(c) $(-101)^{-0}$

4 If you write a letter to a friend with your address at the top as follows:

Please make four identical copies of this letter. Return two of the copies to the address at the top and send the other two copies to two friends asking them to do likewise.

As the message moves on how many letters will arrive back at your address?

Perspective

In the previous chapter we discussed three types of number; the natural numbers, the integers and the rational numbers. We displayed their existence by employing invention based upon a pragmatic principle of need or desire to create a consistent mathematical structure. In reality these numbers and their numerals were discovered or invented in a much more haphazard way; they appeared over the centuries – there was no one person or one group of people who set about to invent them. What is clear, however, is that ancient civilizations had a very sophisticated arithmetic and understood many of the aspects of numbers that we have so far discussed. Some 5,500 years ago the ancient Egyptians were etching hieroglyphic numerals on stone. Around 2,000 BC the Babylonians were recording their appreciation of numbers in cuneiform script with wedge shaped symbols etched into clay tablets (**cuneiform** from the Latin **cuneus** meaning **wedge**). By 600 BC the ancient Greeks were starting to look at numbers as abstractions, as entities that are distinct from the numerals used to represent them.

Within the society of ancient Greece, as in some societies even today, the natural numbers were thought to possess mystical and magical properties. They were seen as significant players in the realms of sorcery, prediction and even religion. The numbers 3 and 7 had an especial significance being the representations of perfection; a significance that has passed to us through the centuries – the three wise men, the Holy Trinity, the seventh day as a day of rest and nirvana being represented as the seventh heaven. The number 6 represented perfect harmony because it is both the sum and the product of its proper divisors 1, 2 and 3. The rational numbers were also understood and, because the natural numbers could be thought of as special types of rational numbers, the whole known number system presented a harmony and power that was a reflection of the harmony and power of the gods; that linchpin of the ancient Greek civilization.

This cultural idea of the harmony of a number system based on the natural numbers and their ratios was dealt a severe blow when *Pythagoras* discovered another type of number whose numeral representation could not be written as a ratio of natural numbers. We shall demonstrate the existence of this new type of number by adopting our original approach of invention via need and consider a fifth arithmetic operation of **raising to a power**.

Raising to a power

We defined multiplication as an operation of convenience to avoid having to compute using repetitive addition. However, it soon becomes apparent that multiplication itself is getting out of hand when we are required to multiply two large numbers together. Paradoxically, we find that we can devise a fifth arithmetic operation that will convert multiplication back to addition and thereby make the multiplication of large numbers easier to handle. We start with repetitive multiplication. For example, we denote the repetitive multiplication of four threes as:

$$3 \times 3 \times 3 \times 3 = 3^4$$

and refer to this as **raising 3 to the power 4**. The multiplication of two fives:

$$5 \times 5 = 5^2$$

5 to the power 2 is referred to as **5 squared** because it gives the area of a square of side length 5. For a similar reason the number:

$$2 \times 2 \times 2 = 2^3$$

is referred to as **2 cubed** because it gives the volume of a cube of side length 2. This new operation of raising to a power will now permit us an alternative means of multiplying two numbers together but first we need to construct a table. For example:

Number				Power
9	=	3×3	$= 3^2$	2
27	=	$3 \times 3 \times 3$	$= 3^3$	3
81	=	$3 \times 3 \times 3 \times 3$	$= 3^4$	4
243	=	$3 \times 3 \times 3 \times 3 \times 3$	$= 3^5$	5

$$792 = 3 \times 3 \times 3 \times 3 \times 3 \times 3 \qquad = 3^6 \qquad 6$$
$$2187 = 3 \times 3 \times 3 \times 3 \times 3 \times 3 \times 3 \qquad = 3^7 \qquad 7$$

The multiplication of a pair of numbers in the left-hand column can now be found by adding the appropriate powers in the right-hand column. For example:

$27 \times 81 = 3^3 \times 3^4$ comparing left- and right-hand sides of the table

$\qquad = (3 \times 3 \times 3) \times (3 \times 3 \times 3 \times 3)$ by the definition of raising to a power

$\qquad = 3 \times 3 \times 3 \times 3 \times 3 \times 3 \times 3$ removing the brackets

$\qquad = 3^7$ by the definition of raising to a power

$\qquad = 2187$ comparing right- and left-hand sides of the table

Notice that:

$$3^7 = 3^{3+4}$$

that is:

$$3^3 \times 3^4 = 3^{3+4}$$

Provided we have the table we can now multiply any pair of these numbers by:

■ *Looking up the powers appropriate to the numbers to be multiplied*
■ *Adding the powers*
■ *Looking up the number corresponding to the sum of the powers*

By converting the process of multiplication to that of addition we have simplified the arithmetic manipulation because the process of addition is a simpler process than that of multiplication.

Before we consider further aspects of constructing the table we need to look a little more closely at what we have invented because there is an implied restriction imposed upon it.

Every power must be a natural number greater than or equal to 2.

We do not, as yet, know what meaning, if any, can be ascribed to numeral combinations of number and power such as:

$5^1, 3^0, 7^{-4}$ or $6^{3/4}$

Worked Examples

2.1 By considering powers of 4 demonstrate how you could, by adding powers, show that:

$$16 \times 64 = 1024$$

Solution:
If you were to construct a table of powers of 4 you would find that:

$$16 = 4^2$$
$$64 = 4^3 \text{ and}$$
$$1024 = 4^5$$

so that:

$$16 \times 64 = 4^2 \times 4^3 = 4^{2+3} = 4^5 = 1024$$

2.2 The sum of the successive odd numbers is always a perfect square. That is, for example:

$$
\begin{aligned}
1 && &= 1^2 \\
1 + 3 && = 4 &= 2^2 \\
1 + 3 + 5 && = 9 &= 3^2 \\
1 + 3 + 5 + 7 && = 16 &= 4^2
\end{aligned}
$$

(a) Construct a geometric demonstration of the validity of this statement.
(b) What is the sum of all the odd numbers up to 21?

Solution:
(a) Construct a geometric demonstration of the validity of this statement.

From the Figure we can see that to increase the square side by one dot requires the next odd number of dots be incorporated into the Figure.

(b) What is the sum of all the odd numbers up to 21?

21 is the 11th odd number and the square of 11 is 121

Exercises

2.1 By considering powers of 2 demonstrate how you could, by adding powers, show that:

$8 \times 16 = 128$

2.2 The sum of the successive cubed integers is always a perfect square. That is, for example:

$$1^3 \qquad\qquad\qquad\qquad = 1^2$$
$$1^3 + 2^3 \qquad\qquad = 9 \quad = 3^2$$
$$1^3 + 2^3 + 3^3 \qquad = 36 \quad = 6^2$$
$$1^3 + 2^3 + 3^3 + 4^3 = 100 = 10^2$$

Deduce the sum of the first ten cubed integers.

Integer powers

As we have pointed out before, restricting operations between numbers to specific numbers is unsatisfactory and provides a stimulus for extending the definition of the operation. By a straightforward extension of a pattern we can define the meaning of raising a number to any integer power as follows:

Number		Power	
$16 = 2 \times 2 \times 2 \times 2$	$= 2^4$	4	
$8 = 2 \times 2 \times 2$	$= 2^3$	3	the power decreases by 1 to 3
$4 = 2 \times 2$	$= 2^2$	2	the power decreases by 1 to 2

In each successive row of this table we are dividing the number in the previous row by 2, thereby effecting a corresponding decrease in the power by 1. At this point in the table we *extend* the definition of raising to a power by continuing to decrease the power by 1 for each successive division:

2	$= 2^1$	1	decreasing the power by 1 to 1
1	$= 2^0$	0	decreasing the power by 1 to 0
$1/2 = 1/(2^1)$	$= 2^{-1}$	-1	decreasing the power by 1 to -1
$1/4 = 1/(2^2)$	$= 2^{-2}$	-2	decreasing the power by 1 to -2
$1/8 = 1/(2^3)$	$= 2^{-3}$	-3	decreasing the power by 1 to -3
$1/16 = 1/(2^4)$	$= 2^{-4}$	-4	decreasing the power by 1 to -4

By successively dividing the previous number by 2 and extending the power column in a natural manner as a sequence of decreasing integers the pattern dictates that we define, for example:

$$2^1 = 2, 2^0 = 1 \text{ and } 2^{-3} = 1/2^3$$

Indeed:

- ■ *raising a number to the power 1 leaves the number unchanged.*
- ■ *any number raised to the power 0 is defined to be 1*
- ■ *negative powers indicate a reciprocal*

Worked Examples

2.3 Find the value of each of the following:
(a) 3^5 (b) 5^{-4}
(c) $(-325)^0$

Solution:
(a) $3^5 = 3 \times 3 \times 3 \times 3 \times 3 = 273$
(b) $5^{-4} = 1/5^4 = 1/[5 \times 5 \times 5 \times 5] = 1/625$
(c) $(-325)^0 = 1$

2.4 There are 64 squares on a chessboard. One grain of rice is placed on the first square, 2 grains of rice are placed on the on the second square, 4 grains on the third square and so on, doubling up until each square has grains of rice on it. How many grains of rice would there be on:
(a) the last square?   (b) the entire board?

Solution:
(a) the last square?

The following table gives the numbers of grains per square:

Square	Grains
1	$1 = 2^0$
2	$2 = 2^1$
3	$4 = 2^2$
4	$8 = 2^3$
..	....
64	$? = 2^{63}$

$2^{63} = 9223372036854775808$ – a very large number.

(b) the entire board?

On the entire board there are:

$$2^0 + 2^1 + 2^2 + 2^3 + ... + 2^{63} = 18446744073709551615$$

an incredible chessboard!

Exercises

2.3 Find the value of each of the following:
(a) 8^4 (b) 9^{-2}
(c) $(-674)^0$

2.4 A mathematical tree consists of nodes and branches where each branch
starts and ends at a node. It is possible to travel through the tree by moving
from node to node along the branches. In any journey through the tree it
is not possible to visit any node more than once and there is only one node,
referred to as the tree root, from which all other nodes can be reached. A
binary tree has at most two branches starting at any node and a complete
binary tree has exactly two branches starting from all nodes except the outer
nodes.

(a) How many different paths are there from the root to the outer nodes
in a three level complete binary tree?
(b) How many different paths are there from the root to the outer nodes
in a four level complete binary tree?
(c) How many different paths are there from the root to the outer nodes
in a twenty level complete binary tree?

Unit 2   Rational powers

Try the following test:

1   Find the value of each of the following:
(a) $2^3 \times 2^6$ (b) $5^4 \times 5^{-3}$
(c) $(3^4)^2$ (d) $(-4^2) \times 2^2$

2   Find the value of each of the following (taking positive roots where appropriate):
(a) $(16)^{-5/4}$ (b) $4^{7/4} \times 4^{-1/4}$
(c) $(81^{1/4})^4$ (d) $(1/5)^{3/2} \times 5^{-1/4}$
(e) $((1/7)^{1/3})^{-3}$

3   Simplify each of the following (taking positive roots where appropriate):
(a) $[6^2 \times 36^3 - 12^2 \times 6^1]/6^3$
(b) $[(5^{-2}) \times (25^{1/2}) - (125^{2/3}) \times (5^{-1})]/[25^{-2} \times 5^3 + 5^4 \times 5^{-3}]$

4   Show that:
(a) $\sqrt{27} = 3\sqrt{3}$
(b) $\sqrt{2} + 1/\sqrt{2} = 3/\sqrt{2}$
(c) $[\sqrt{125} + \sqrt{5}]/[\sqrt{125} - \sqrt{5}] = 3/2$
(d) $[\sqrt{18} + \sqrt{6}]/[\sqrt{12} + 1] = \sqrt{(3/2)}$

5   Many rational numbers are the square of an irrational number. Is it possible to have an irrational number that is the square of a rational number?

The arithmetic of powers
Having extended our mathematical structure to include a fifth operation – raising a number to an integer power – we must now look to the rules that govern the use of this operation:

Addition of powers
Powers can be added to effect a multiplication:

$$9^3 \times 9^4 = 9^7 = 9^{3+4}$$

Subtraction of powers
Powers can be subtracted to effect a division:

$$9^3 \div 9^4 = 9^3 \times (1/9^4)$$
$$= 9^{3-4}$$
$$= 9^{-1}$$

division is multiplication by the reciprocal

Multiplication of powers
Powers can be multiplied to effect a raising to a power:

$$(9^3)^4 = (9^3) \times (9^3) \times (9^3) \times (9^3)$$
$$= 9^3 \times 9^3 \times 9^3 \times 9^3$$
$$= 9^{12}$$
$$= 9^{3 \times 4}$$

Division of powers
The division of powers is a problem. For example:

$9^{3/4}$ can be written as $(9^3)^{1/4}$ or $(9^{1/4})^3$

But in either case we do not know what is meant by raising a number to a fractional power. We must attack this problem.

Worked Examples

2.5 Find the value of each of the following:
    (a) $3^4 \times 3^2$ (b) $7^2 \times 7^{-3}$
    (c) $(5^2)^3$ (d) $(-2^4) \times (8^4)$

Solution:
(a) $3^4 \times 3^2 = 3^{4+2} = 3^6 = 729$
(b) $7^2 \times 7^{-3} = 7^{2-3} = 7^{-1} = 1/7$
(c) $(5^2)^3 = 5^{2 \times 3} = 5^6 = 15625$
(d) $(-2^4) \times (8^4) = (-16)^4 = 65536$

Notice from (d) that if two different numbers are both raised to the *same* power then we can write this product as the product of the two numbers raised to the power. We cannot do this if two different numbers are raised to different powers.

Exercises

2.5 Find the value of each of the following:
    (a) $4^2 \times 4^5$ (b) $6^5 \times 6^{-2}$
    (c) $(8^3)^3$ (d) $(-5^2) \times (3^2)$

Rational powers

We understand what is meant by an integer power and we know how to perform the arithmetic of integer powers. We do not as yet understand what is meant by a rational power and in particular we do not as yet know the meaning of, for example:

$$16^{1/4}$$

We do know how to add powers so we can say that:

$$16^{1/4} \times 16^{1/4} \times 16^{1/4} \times 16^{1/4} = 16^{1/4 + 1/4 + 1/4 + 1/4}$$
$$= 16^{1}$$
$$= 16$$

That is, when four identical numbers, each represented by $16^{1/4}$ are multiplied together the result equals 16. We call $16^{1/4}$ the **fourth root** of 16 and, because:

$$2 \times 2 \times 2 \times 2 = 16$$

we can see that:

$$16^{1/4} = 2$$

Again we have two exceptions to the terminology:

$$9^{1/2}$$

the second root of 9, which is 3, is called the **square root** of 9 because it can represent the side length of a square whose area is 9. Also:

$$125^{1/3}$$

the third root of 125, which is 5, is called the **cube root** of 125 because it can represent the side length of a cube whose volume is 125.

Even roots

The even root of a number is not unique. For example, because:

$$4 \times 4 = 16 \text{ and } (-4) \times (-4) = 16$$
we see that

$$16^{1/2} = 4 \text{ } or -4$$

We write this as:

$$16^{1/2} = \pm 4 \text{ where the symbol } \pm \text{ stands for } \textbf{plus or minus.}$$

Worked Examples

2.6 Find the value of each of the following (taking positive roots where
appropriate):
(a) $8^{-2/3}$ (b) $(27)^{4/3}$
(c) $3^{2/3} \times 3^{1/2}$ (d) $7^{5/6} \times 7^{-1/3}$
(e) $(5^{1/2})^2$ (f) $(-2^{1/4}) \times (-8^{1/4})$
(g) $(1/3)^{-2/3} \times 3^{-1/2}$ (h) $7^{5/6} \times (1/7)^{-1/3}$
(j) $((1/5)^{1/2})^{-2}$ (k) $(-1/2^{1/2}) \times (-8^{1/2})$

Solution:
(a) $8^{-2/3} = ((8)^{1/3})^{-2} = (2)^{-2} = 1/4 \ or \ 8^{-2/3} = ((8)^{-2})^{1/3} = (1/64)^{1/3} = 1/4$
(b) $(27)^{4/3} = (27^{1/3})^4 = 3^4 = 81$
(c) $3^{2/3} \times 3^{1/2} = 3^{2/3 + 1/2} = 3^{4/6 + 3/6} = 3^{7/6}$
(d) $7^{5/6} \times 7^{-1/3} = 7^{5/6 - 1/3} = 7^{5/6 - 2/6} = 7^{3/6} = 7^{1/2}$
(e) $(5^{1/2})^2 = 5^{1/2 \times 2} = 5^1 = 5$
(f) $(-2^{1/4}) \times (-8^{1/4}) = 16^{1/4} = 2$
(g) $(1/3)^{-2/3} \times 3^{-1/2} = (3^{-1})^{-2/3} \times 3^{-1/2} = 3^{2/3} \times 3^{-1/2} = 3^{2/3 - 1/2} = 3^{1/6}$
(h) $7^{5/6} \times (1/7)^{-1/3} = 7^{5/6} \times 7^{1/3} = 7^{7/6}$
(j) $((1/5)^{1/2})^{-2} = (1/5)^{-1} = 5$
(k) $(-1/2^{1/2}) \times (-8^{1/2}) = 4^{1/2} = 2$

Notice that in (f) and (k) whilst we have not defined the even root of a negative
number we can answer these questions only because the product of two different
numbers, each raised to the same power is equal to their product raised to that power.

2.7 Simplify each of the following (taking positive roots where appropriate):
(a) $[4^4 \times 8^{-1} - 2^2 \times 2^{-1}]/2$
(b) $[(3^{-1}) \times (9^{3/2}) - (81^{1/2}) \times (9^{-1/2})]/[3^{-1} \times 3^3 + 3^2 \times 3^{-1}]$

Solution:
(a) $[4^4 \times 8^{-1} - 2^2 \times 2^{-1}]/2 = [256/8 - 4/2]/2 = [32 - 2]/2 = 30/2 = 15$

(b) $[(3^{-1}) \times (9^{3/2}) - (81^{1/2}) \times (9^{-1/2})]/[3^{-1} \times 3^3 + 3^2 \times 3^{-1}]$
$= [(1/3) \times (3^3) - (9) \times (1/3)]/[3^2 + 3^1] = [9 - 3]/[9 + 3] = 6/12 = 1/2$

Exercises

2.6 Find the value of each of the following (taking positive roots where
appropriate):
(a) $9^{3/2}$ (b) $(-16)^{-5/4}$
(c) $5^{3/4} \times 5^{1/5}$ (d) $4^{7/8} \times 4^{-1/4}$
(e) $(16^{1/3})^3$ (f) $(-5^{1/3}) \times (-25^{1/3})$
(g) $(1/7)^{-3/4} \times 7^{-2/3}$ (h) $4^{2/3} \times (1/4)^{-1/4}$
(j) $((1/3)^{1/5})^{-5}$ (k) $(-1/3^{1/2}) \times (-27^{1/2})$

2.7 Simplify each of the following (taking positive roots where appropriate):
  (a) $[5^3 \times 10^2 - 25^2 \times 5^1]/5^2$
  (b) $[(2^{-1}) \times (4^{3/2}) - (64^{1/2}) \times (16^{-1/2})]/[4^{-1} \times 4^3 + 4^2 \times 4^{-1}]$

Retrospect
We have come a long way since those early days when we wanted to make a permanent record of our counting by scratching symbols on the walls of our cave. From an immediate desire to count we have, over time, developed an abstract mathematical structure consisting of:

■ *Rational numbers represented as ratios of integers where each integer, being positive or negative, can be written using combinations of the ten numerals located in place value order.*

■ *Five operations that act on the rational numbers:*

> **addition**
> **subtraction**
> **multiplication**
> **division**
> **raising to a power**

where it is assumed that the effect of combining rational numbers under all these operations produces another rational number. In addition, the precedence rules for combining rational numbers follow the BEODMAS rules where the E stands for **exponentiation** or raising to a power.

The only restrictions that have been placed on the structure so far are that we cannot define the largest positive or smallest negative integer and we cannot define division by zero. Since both of these restrictions are interrelated and cannot be resolved we must learn to live with them.

Such was the happy state of affairs until *Pythagoras* and his followers made a startling discovery.

Irrational numbers
The number represented by:

$$2^{1/2}$$

cannot also be represented as a ratio of integers. In other words the number represented by $2^{1/2}$ is **not a rational number**. The disturbance that this discovery created amongst the Pythagoreans is explained by the fact that it is a number that stands outside the comfortable, harmonious structure that reflected the philosophy of the time.

It would be instructive to see a proof of this statement at this stage were it not that the proof requires an ability to manipulate algebraic symbols – the subject matter of the second part of this book. As a consequence the proof has been assigned to the Appendix at the end of the book where you can peruse it at leisure. For now it is sufficient if you will accept the fact that:

$2^{1/2}$ is not a rational number

If it is not a rational number then what sort of number is it? It is a perfectly bona fide number because if you multiply it by itself you do get a rational number; you get a natural number in fact, the number 2.

When this fact was first discovered by *Pythagoras* and his adherents it created a deal of consternation, caused in the main by the fact that there was no distinction made between a number and the numeral used to represent it. Now we know better and when we say that the square root of two is not a rational number we merely mean that it cannot be represented by a ratio of integers. Such a number is called an **irrational** number. The use of the word irrational is a rather unfortunate choice because there is nothing behaviourally irrational about such numbers, only the fact that they cannot be represented by ratios of integers. Because irrational numbers cannot be represented by ratios of integers other methods of representing them have been invented over time. For example, the irrational numbers:

$3^{1/2}$ and $5^{1/3}$

can be alternatively represented by the notation ($\sqrt{}$):

$\sqrt{3}$ and $^3\sqrt{5}$ respectively.

Some irrational numbers cannot be expressed as a rational number raised to a power so other symnbols are used. For example, the irrational number that represents the proportionality constant linking the circumference to the ratio of a circle is represented by the symbol:

π

Pronounced **pi**, this symbol is a lower-case Greek letter **p**.

Irrational numbers are far more numerous than rational numbers. Indeed, there are more irrational numbers between the integers 0 and 1 than there are rational numbers in their entirety. Whilst it is inappropriate in this book to attempt to prove this statement it is mentioned just in case you are tempted to think of irrational numbers as some sort of rare oddity.

The real numbers

The totality of all the rational numbers and all the irrational numbers forms the **real** numbers. Again the word real connotes some concrete existence which they do not possess and as such is an unfortunate choice of word.

The real numbers are ordered and are represented graphically as points on a line where the direction of increasing number is indicated by the arrow. Notice that:

5 > 3 because the point 5 is to the right of the point 3. For the same reason,

$0 > -1$ and $-5 < -3$

The line on which these points are located is called the **real line** and it extends indefinitely in either direction; there is no largest (positive) and no smallest (negative) real number. Furthermore, the line is complete; every point on the line corresponds to a unique real number and to every real number there corresponds a unique point on the line.

Conclusion

We have finally constructed the mathematical structure called the **algebra of the real numbers** or **arithmetic** for short. The structure is complete and consistent. There is, however, one outstanding problem we have not yet mentioned. We do not have a number that corresponds to the even root of a negative number. For example, we do not have a number that corresponds to:

$$(-1)^{1/2} \quad \text{where} \quad (-1)^{1/2} \times (-1)^{1/2} = (-1)^1 = -1$$

Whatever the number represented by the numeral $(-1)^{1/2}$ is, it **cannot be a real number** for two reasons. Firstly, real numbers are positive, negative or zero and when we multiply two positive numbers or two negative numbers we always end up with a positive number. We do not have a real number which, when multiplied by itself gives -1. Secondly, the real line is complete – there is no unplotted point remaining where we could plot this number.

The number that $(-1)^{1/2}$ represents, however, must be a bona fide number because when multiplied by itself it gives the real number -1. To cater for this problem we are going to have to put on our pragmatic hat once more – but that is another story . . .

Worked Examples

2.8 Show that:
    (a) $1/\sqrt{2} = \sqrt{2}/2$
    (b) $\sqrt{2} - 1/\sqrt{2} = 1/\sqrt{2}$
    (c) $[\sqrt{8} - \sqrt{2}]/[\sqrt{8} + \sqrt{2}] = 1/3$
    (d) $1/[\sqrt{6} + \sqrt{3}] = (\sqrt{2} - 1)/\sqrt{3}$

Solution:
(a) $1/\sqrt{2} = \sqrt{2}/2$

    $2 = \sqrt{2} \times \sqrt{2}$ so that by dividing both sides by $\sqrt{2}$ we see that:
    $2/\sqrt{2} = \sqrt{2}$ and hence, by taking the reciprocal of both sides that:
    $\sqrt{2}/2 = 1/\sqrt{2}$

(b) $\sqrt{2} - 1/\sqrt{2} = 1/\sqrt{2}$

    $\sqrt{2} - 1/\sqrt{2} = (\sqrt{2} \times \sqrt{2})/\sqrt{2} - 1/\sqrt{2}$

$$=2/\sqrt{2} - 1/\sqrt{2}$$
$$= (2 - 1)/\sqrt{2}$$
$$= 1/\sqrt{2}$$

(c) $[\sqrt{8} - \sqrt{2}]/[\sqrt{8} + \sqrt{2}] = 1/3$

$$[\sqrt{8} - \sqrt{2}]/[\sqrt{8} + \sqrt{2}] = [2\sqrt{2} - \sqrt{2}]/[2\sqrt{2} + \sqrt{2}]$$

because: $\sqrt{8} = \sqrt{(4 \times 2)}$
$$= \sqrt{4} \times \sqrt{2}$$
$$= 2\sqrt{2}$$

Hence:

$$[\sqrt{8} - \sqrt{2}]/[\sqrt{8} + \sqrt{2}] = \sqrt{2}/3\sqrt{2}$$
$$= 1/3$$

(d) $1/[\sqrt{6} + \sqrt{3}] = (\sqrt{2} - 1)/\sqrt{3}$

$$1/[\sqrt{6} + \sqrt{3}] = ([\sqrt{6} - \sqrt{3}]/[\sqrt{6} - \sqrt{3}]) \times (1/[\sqrt{6} + \sqrt{3}])$$
$$= [\sqrt{6} - \sqrt{3}]/([\sqrt{6} - \sqrt{3}] \times [\sqrt{6} + \sqrt{3}])$$
$$= [\sqrt{6} - \sqrt{3}]/(\sqrt{6} \times [\sqrt{6} + \sqrt{3}] - \sqrt{3} \times [\sqrt{6} + \sqrt{3}])$$
$$= [\sqrt{6} - \sqrt{3}]/(\sqrt{6} \times \sqrt{6} + \sqrt{6} \times \sqrt{3} - (\sqrt{3} \times \sqrt{6} + \sqrt{3} \times \sqrt{3}))$$
$$= [\sqrt{2}\sqrt{3} - \sqrt{3}]/[6 + \sqrt{18} - \sqrt{18} - 3]$$
$$= (\sqrt{2} - 1)\sqrt{3}/3$$
$$= (\sqrt{2} - 1)/\sqrt{3}$$

2.9 The ratio of the circumference to the diameter of a circle is given as the irrational quantity π. Given that an irrational number is not expressible as a ratio of integers what does this tell you about the circumference and the diameter of a circle?

Solution:
The length of the diameter and the length of the circumference cannot both be represented by a rational number. Either one or both must be represented by an irrational number.

Exercises

2.8 Show that:
   (a) $\sqrt{8} = 2\sqrt{2}$ (b) $\sqrt{3} + 1/\sqrt{3} = 4/\sqrt{3}$
   (c) $[\sqrt{27} + \sqrt{3}]/[\sqrt{27} - \sqrt{3}] = 5/4$ (d) $[\sqrt{3} - 1]/[\sqrt{15} - \sqrt{5}] = \sqrt{5}/5$

2.9 Given a square of area 2 square units what can you say about the length of a side?

Module 2 Further exercises

1 By considering powers of 5 demonstrate how you could, by adding powers, show that:

$$25 \times 125 = 3125$$

2 Extend the following pair of equations and find out if there is a pattern:

$$3^2 + 4^2 \quad = 5^2$$
$$3^3 + 4^3 + 4^3 = 6^3$$

3 Find the value of each of the following:
  (a) 6^5　　　　　　　　　　　　　　(b) 4^{-3}
  (c) $(-32)^0$

4 Complete the following pattern of equations and then extend it by two rows back and two rows forward:

$$3^2 + 4^2 + 12^2 = \blacksquare$$
$$4^2 + 5^2 + 20^2 = \blacksquare$$
$$5^2 + \blacksquare + 30^2 = 31^2$$

5 Find the value of each of the following:
  (a) $5^2 \times 5^3$　　　　　　　　　　(b) $4^5 \times 4^{-2}$
  (c) $(3^2)^3$　　　　　　　　　　　　(d) $(-4^3) \times (6^3)$

6 Find the value of each of the following (taking positive roots where appropriate):
  (a) $25^{3/2}$　　　　　　　　　　　　(b) $4^{2/3} \times 4^{1/4}$
  (c) $(8^{2/3})^{3/2}$　　　　　　　　　　(d) $(1/6)^{-3/5} \times 6^{-5/3}$
  (e) $((1/8)^{-1/7})^{-7}$

7 Simplify each of the following (taking positive roots where appropriate):
  (a) $[3^4 \times 9^3 - 6^4 \times 3^2]/3^3$
  (b) $[(2^{-3}) \times (4^{2/3}) - (8^{1/3}) \times (16^{3/2})]/[4^{-2} \times 2^3 + 8^2 \times 4^2]$

8 Show that:
  (a) $\sqrt{125} = 5\sqrt{5}$　　　　　　　(b) $4/\sqrt{5} + \sqrt{5}/4 = 21/4\sqrt{5}$
  (c) $[\sqrt{32} - \sqrt{8}]/[\sqrt{32} + \sqrt{8}] = 1/3$　　(d) $[\sqrt{27} + \sqrt{8}]/\sqrt{6} = 3/\sqrt{2} + 2/\sqrt{3}$

9 Given a square of area 15 square units what can you say about the length of a side?

Module 3

Decimals

OBJECTIVES

When you have completed this module you will be able to:

■ Manipulate the arithmetic of decimal numbers both by hand and by calculator

---

There are three units in this module:

Unit 1: The number base 10
Unit 2: Truncating decimal numbers
Unit 3: The arithmetic of decimals

Unit 1   The number base 10

Try the following test:

1   Write each of the following numbers in the form of sum of coefficients multiplying 10 to the appropriate power:
   (a) 63052   (b) −324
   (c) 31.25   (d) 80.008

2   By finding equivalent fractions convert each of the following fractions to decimal numbers:
   (a) 2/5   (b) 3/25
   (c) 17/4   (d) −7/8

3   Write each of the following decimal numbers in a shorthand form:
   (a) $5/7 = 0.714285714285714285...$
   (b) $3/11 = 0.272727...$
   (c) $-5/6 = -0.83333333...$
   (d) $-7/41 = -0.107073107073107073...$

Place value

At the very beginning of our development of the real number system it was mentioned that the integers are written down using the ten numerals located in place value order. We stated that in the following arrangement of numerals:

32425

the location of a numeral within the arrangement dictates the number that it represents. For instance, the numeral 2 to the left of the numeral 4 represents the number 2000 whereas the numeral 2 to the right of the numeral 4 represents the number 20. Indeed, the number represented by the numeral arrangement 32425 can be alternatively represented by the sum:

30000 + 2000 + 400 + 20 + 5

Noting that:

$$10000 = 10^4$$
$$1000 = 10^3$$
$$100 = 10^2$$
$$10 = 10^1$$
$$1 = 10^0$$

we can represent the sum:

$$30000 + 2000 + 400 + 20 + 5$$

as the sum:

$$3 \times 10^4 + 2 \times 10^3 + 4 \times 10^2 + 2 \times 10^1 + 5 \times 10^0$$

By writing this last sum in an array of columns, each headed by 10 raised to the appropriate power, we retrieve the first representation of the number:

10^4	10^3	10^2	10^1	10^0
3	2	4	2	5

The numerals 3, 2, 4, 2 and 5 are referred to as the **coefficients** of the respective powers of 10. By writing the coefficients in descending order of powers in this way we obtain a number written with numerals in place value order; each place, or column, is assumed to be headed by a power of 10 which determines the number represented by a numeral placed in that column.

Decimal numbers and decimal arithmetic

By extending the range of columns to include higher powers of 10 it is possible to write any integer using numerals in place value order. By extending the columns to include all integer powers – negative as well as positive – we find that we can also express ratios of integers using place value order provided that we indicate those columns that refer to negative powers of 10.

For example, by noting that:

$$1/2 = 5/10$$
$$= 5 \times 10^{-1}$$

we can write this number using the numeral 5 in place value order as:

$$.5$$

where the point – the **decimal point** – indicates that the 5 is the coefficient of the first negative power of 10. To make this distinction even clearer it is good practice to write the number as:

$$0.5$$

where the zero emphasizes the presence of the decimal point. The invention of the decimal point permits us to distinguish between two different numbers when each number is represented by the same list of numerals. For example, the list of numerals:

$$32425$$

represents a different number from the list of numerals:

3242.5

This latter number is:

$$3 \times 10^3 + 2 \times 10^2 + 4 \times 10^1 + 2 \times 10^0 + 5 \times 10^{-1} = 3242 + 5/10$$

The decimal point has indicated where the negative powers of 10 begin. We refer to this form of representation of a number as the **decimal representation** (from the Latin **decem** meaning **ten**). By the same reasoning the following is true:

$$324.25 = 3 \times 10^2 + 2 \times 10^1 + 4 \times 10^0 + 2 \times 10^{-1} + 5 \times 10^{-2}$$
$$= 324 + 2 \times 10^{-1} + 5 \times 10^{-2}$$
$$= 324 + 2/10 + 5/100$$

$$32.425 = 3 \times 10^1 + 2 \times 10^0 + 4 \times 10^{-1} + 2 \times 10^{-2} + 5 \times 10^{-3}$$
$$= 32 + 4 \times 10^{-1} + 2 \times 10^{-2} + 5 \times 10^{-3}$$
$$= 32 + 4/10 + 2/100 + 5/1000$$

and:

$$3.2425 = 3 \times 10^0 + 2 \times 10^{-1} + 4 \times 10^{-2} + 2 \times 10^{-3} + 5 \times 10^{-4}$$
$$= 3 + 2 \times 10^{-1} + 4 \times 10^{-2} + 2 \times 10^{-3} + 5 \times 10^{-4}$$
$$= 3 + 2/10 + 4/100 + 2/1000 + 5/10000$$

or, to use the column display:

10^2	10^1	10^0		10^{-1}	10^{-2}	10^{-3}	10^{-2}
3	2	4	.	2	5		
	3	2	.	4	2	5	
		3	.	2	4	2	5

Strictly speaking, we should enumerate the coefficients of every power of 10 but in reality this is not practicable. Take for example the number:

$$... + 0 \times 10^2 + 0 \times 10^1 + 5 \times 10^0 + 2 \times 10^{-1} + 3 \times 10^{-2} + 0 \times 10^{-3} + 0 \times 10^{-4} + ...$$

In the decimal representation this should be written as:

...005.2300...

To make sense of the decimal representation we truncate the representation and:

omit all leading and all trailing zeros

We write the number as 5.23

Worked Examples

3.1 Write each of the following numbers in the form of sum of coefficients multiplying 10 to the appropriate power:
(a) 75342   (b) −265
(c) 529.76   (d) 100.001

Solution:
(a) $75342 = 70000 + 5000 = 300 + 40 + 2$
$= 7 \times 10^4 + 5 \times 10^3 + 3 \times 10^2 + 4 \times 10^1 + 2 \times 10^0$

(b) $-265 = -(200 + 50 + 6)$
$= -(2 \times 10^2 + 6 \times 10^1 + 5 \times 10^0)$
$= -2 \times 10^2 - 6 \times 10^1 - 5 \times 10^0$

(c) $529.76 = 500 + 20 + 9 + 7/10 + 6/100$
$= 5 \times 10^2 + 2 \times 10^1 + 9 \times 10^0 + 7 \times 10^{-1} + 6 \times 10^{-2}$
$+ 0 \times 10^{-3} + 0 \times 10^{-4}$

(d) $100.01 = 100 + 1/1000$
$= 1 \times 10^2 + 0 \times 10^1 + 0 \times 10^0 + 0 \times 10^{-1} + 0 \times 10^{-2} + 1$

Exercises

3.1 Write each of the following numbers in the form of a sum of coefficients multiplying 10 to the appropriate power:
(a) 890435   (b) −157
(c) 28.557   (d) 90.004

Rational numbers
The conversion of the representation of a rational number from a ratio of integers to its decimal representation is effected by performing the division. For example, to convert:

3/4

to decimal form we perform the division:

```
        0.75
  4 ⌐ 3.00
      2.80
       .20
       .20
         0
```

So that 3/4 = 0.75

For some rational numbers we find that repeated division produces a decimal of infinite extent. For example the rational number represented by:

1/3

has a decimal representation that is obtained by division as:

0.3333333...

Here the 3 repeats itself indefinitely. We agree to write this as:

$0.\dot{3}$

where the dot above the 3 tells us that the numeral 3 is repeated indefinitely. The number 1/7 has the decimal representation:

1/7 = 0.142857142857142857...

Here the sequence of numerals 142857 repeats itself indefinitely. We write

$0.\dot{1}4285\dot{7}$

where the dots above the 1 and 7 tell us that the sequence of numerals is repeated indefinitely.

Worked Examples

3.2 By finding equivalent fractions convert each of the following fractions to decimal numbers:
(a) 1/4
(b) 3/4
(c) 15/2
(d) −3/5

Solution:
(a) 1/4 = 25/100 = 0.25

(b) 3/4 = 1/2 + 1/4 = 5/10 + 25/100 = 50/100 + 25/100 = 75/100 = 0.75

(c) 15/2 = 7 + 1/2 = 7 + 5/10 = 7.5

(d) −3/5 = −3 × (1/5) = −3 × (2/10) = −6/10 = −0.6

3.3 Write each of the following decimal numbers in a shorthand form:

    (a) 2/7 = 0.285714285714285...   (b) 1/13 = 0.0762930762930762...

    (c) −5/6 = −0.83333333...   (d) −5/33 = −0.151515...

Solution:

(a) 2/7 = 0.285714285714285714...

$$= 0.28571\dot{4}$$

(actually: $= 0.\dot{2}8571\dot{4}$)

(b) 1/13 = 0.076293076293076293...

$$= 0.\dot{0}7629\dot{3}$$

(c) −5/6 = −0.83333333...

$$= −0.8\dot{3}$$

(d) −5/33 = −0.151515...

$$= −0.\dot{1}\dot{5}$$

Exercises

3.2 By finding equivalent fractions convert each of the following fractions to decimal numbers:

    (a) 5/20   (b) 7/4

    (c) 23/2   (d) −12/15

3.3 Write each of the following decimal numbers in a shorthand form:

    (a) 3/7 = 0.4285714285714285...   (b) 2/11 = 0.181818...

    (c) −7/12 = −0.583333333...   (d) −4/37 = −0.108108108...

Unit 2　　Truncating decimal numbers

Try the following test:

1　In each of the following identify the 1000th numeral after the decimal point:
　(a)　5/7 = 0.7142857142857142...
　(b)　3/11 = 0.272727...
　(c)　−5/6 = −0.83333333...
　(d)　−7/41 = −0.107073107073107073...

2　Truncate each of the following decimal numbers to one, two and three decimal places:
　(a)　25.45273　　　　　　　　(b)　−0.027561
　(c)　19.191919　　　　　　　　(d)　−16.0548

3　Truncate each of the following decimal numbers to three, four and five significant figures:
　(a)　25.45273　　　　　　　　(b)　−0.027561
　(c)　19.191919　　　　　　　　(d)　−16.0548

4　Write each of the following numbers in scientific notation:
　(a)　0.00000632　　　　　　　(b)　930000
　(c)　300054　　　　　　　　　(d)　−0.00203

5　Round each of the following decimal numbers to the nearest tenth, unit and thousand:
　(a)　26102.837　　　　　　　　(b)　−794.329
　(c)　1655.55　　　　　　　　　(d)　−1099.99

6　Find two different decimal representations of each of the following numbers:
　(a)　2/5　　　　　　　　　　　(b)　−1/2
　(c)　7/8　　　　　　　　　　　(d)　−9/25

Irrational numbers

Every irrational number has a decimal representation that is of infinite extent with no sequence of numerals that repeats itself indefinitely. For example, the square root of 2 is given as $\sqrt{2} = 1.4142136...$

Here, the list of numerals after the decimal point is infinite in extent. However, unlike a rational number, it contains no indefinitely repeating sequence of numerals. Now we really do have a problem. Not only can we not write down the decimal representation of an irrational number because it is infinite in extent we cannot even invent a shorthand symbolism that can be taken to represent the decimal form.

   A rational number may have an infinite sequence of numerals after the decimal point but it is always possible to identify the numeral located at any position in the sequence. This is done by implementing the pattern of repeated numerals and performing a simple count. For example:

   the numeral in the 100th position after the decimal point in the decimal representation of 1/7 is 8 because the first 96 locations are occupied by the numerals 142857 repeated 16 times and the 100th location is the 4th numeral in the 17th repetition of the list which is 8.

In the case of an irrational number it is not possible to apply this procedure. Because there is no such repeated pattern involved it is not possible to identify the numeral in any position without a knowledge of all the numerals that come before it. Now that really is a dilemma. By accepting the existence of irrational numbers we assume that they can be quantified – that we can find their exact value. We now find that we cannot quantify them because to do so would require an infinite time to write down their complete decimal representation. We possess a great deal of information about irrational numbers except the one thing that really matters – their exact value.

   We are back to the origins of the number system where we had a desire to record our counting. From this desire we invented counting numbers – the natural numbers. Very soon after this we implemented another desire – the desire to create a fully consistent number system – and that is when our difficulties started to appear. Counting is a process of interaction between the intellect and the physical world. Constructing a consistent number system is a purely intellectual activity that at times may impinge upon the physical world but which usually does so only tangentially. The existence of irrational numbers came about from geometric considerations and the truth is that all the geometric figures with which we are so familar, the point, the line, the triangle, the circle and so on, do not exist as concrete entities, they only exist as figments of the imagination. Certainly we can draw points, lines, triangles and circles but our scratchings are only crude approximations to our intellectual definitions of these figures.

   In conclusion we recognize that quantification is related to concrete entities because quantification implies an ability to measure. We can define the value of the ratio of the circumference to the diameter of a circle as the unquantifiable number π with impunity only because the circle is not a concrete entity.

Worked Examples

3.4 In each of the following identify the 1000th numeral after the decimal point:
   (a) 2/7 = 0.285714285714285714...
   (b) 1/13 = 0.0762930762930762 93...
   (c) –5/6 = –0.83333333...
   (d) –4/9 = –0.090277777...

Solution:
(a) 2/7 = 0.789473684210526315... the 1000th numeral is 2 which is the tenth

numeral in the 56th repetition of the 18 numerals in the repeat.

(b) 1/13 = 0.076293076293076293... the 1000th numeral is 2 which is the fourth numeral in the 167th repetition of the 6 numerals in the repeat.

(c) −5/6 = −0.83333333... the 1000th numeral is 3.

(d) −4/9 = −0.090277777... the 1000th numeral is 7.

Exercises

3.4 In each of the following identify the 1000th numeral after the decimal point:
(a) 3/7 = 0.4285714285714285... (b) 2/11 = 0.181818...
(c) −7/12 = −0.583333333... (d) −4/37 = −0.108108108...

Decimal places and significant figures
A need has been clearly evidenced. We need to be able to write down decimal numbers in a manner that is both convenient and sensible. The numbers that we write down in the decimal representation must reflect their use at the time. For example, whilst it is accepted that:

$$0.\dot{3}$$

is how we describe 1/3 in the decimal representation it is not a very helpful notation for describing 1/3 of a metre to the nearest millimetre. We need a more appropriate method of truncating decimal numbers.

Decimal places
A decimal number can be written to a pre-stated number of decimal places. For example, the number 123.1231211... can be written as:

123.1   to one decimal place
123.12   to two decimal places
123.123   to three decimal places

and so on. The number of decimal places retained depends upon the situation within which the number is being used. For example, if we are using a number to represent the height of an individual then we would measure to the nearest millimetre. In this case a height of 1m 85 cm 4 mm would be denoted as:

1.854 m or 185.4 cm or 1854 mm

Writing a decimal number to a given number of decimal places requires the decimal representation to be truncated. There is a rule about this:

*If the first number to be omitted is less than 5 the last number retained is left
unaltered. Otherwise the last number retained is increased by 1.*

For example, the number 156.754286... written to three decimal places is:

156.754   Here the last number retained, the number 4, is unaltered
                 because the first number omitted is 2 which is less than 5.

Alternatively, if we write this number to four decimal places we write:

156.7543   Here the last number to be retained, the number 2, is increased by
            1 to 3 because the first number omitted is 8 which is not less than 5.

Worked Examples

3.5 Truncate each of the following decimal numbers to one, two and three decimal
places:
   (a) 13.43749   (b) -0.04663
   (c) 1.99999   (d) -25.0106

Solution:
(a) 13.43749: 13.4, 13.44, 13.437

(b) -0.04663: $-0.0 = 0.0$, -0.05, -0.047

(c) 1.99999: 2.0, 2.00, 2.000

(d) -25.0106: -25.0, -25.01, -25.011

Exercises

3.5 Truncate each of the following decimal numbers to one, two and three decimal
places:
   (a) 17.64537   (b) -0.015468
   (c) 7.11111   (d) -36.0909

Writing a number to a specified number of decimal places only takes account of those
numerals to the right of the decimal point. This is not always convenient, especially
if there is a large number of numerals to the immediate left of the decimal point.
Counting to a specified number of **significant figures** is an alternative method of
truncating a decimal number that can also take into account those numerals to the left
of the decimal point.

Significant figures

A decimal number can be written to a pre-stated number of significant figures where the number of significant figures are counted by starting with the first non-zero numeral at the left-hand end of the number. For example, 156.754286... written to five significant figures is:

156.75

Truncating decimal numbers after a given number of significant figures follows the same rule as that previously discussed. For example, this number written to four significant figures is:

156.8   Notice that the first number omitted is 5 which is not less than 5, hence the last retained figure 7 is increased by 1 to 8.

When counting significant figures the first non-zero numeral from the left is the first significant figure. For example, the truncated form of 0.00000135 to two significant figures is:

0.0000014

Similarly, the number 123456789 written to four significant figures is:

123500000

Worked Examples

3.6 Truncate each of the following decimal numbers to three, four and five significant figures:
(a) 13.43749
(b) −0.0466355
(c) 1.99991
(d) −25.0106

Solution:
(a) 13.43749: 13.4, 13.44, 13.437

(b) −0.0466355: −0.0466, −0.04664, −0.046636

(c) 1.99991: 2.00, 2.000, 1.9999

(d) −25.0106: −25.0, −25.01, −25.011

Exercises

3.6 Truncate each of the following decimal numbers to three, four and five significant figures:

(a) 17.64537   (b) −0.015468
(c) 7.11111   (d) −36.0909

Numerals where a large number of zeros lie between the decimal point and the first or last significant figure clearly pose a problem even when truncated. We need a better way of representing truncated decimal numbers.

Scientific notation
The scientific notation of a decimal number was devised to eliminate a large number of leading or trailing zeros when writing down a decimal number to a given number of significant figures. This involves rewriting the number as a number between 1 and 10 (if the number is positive, otherwise between −1 and −10) multiplied by 10 raised to the appropriate power. For example, the number 0.0000014 can be written as the sum:

$$1 \times 10^{-6} + 4 \times 10^{-7} = (1 + 4 \times 10^{-1}) \times 10^{-6}$$

That is:

$$1.4 \times 10^{-6}$$

This latter expression is the **scientific notation** form of the number 0.0000014. Written in this form the number:

    10 is referred to as the **base**
    −6 is referred to as the **exponent**.
    1.4 is referred to as the **mantissa**

Again, by the same reasoning the number 123500000 is written in scientific notation as:

$$1.235 \times 10^{8}$$

Worked Examples

3.7 Write each of the following numbers in scientific notation:
(a) 0.0000567   (b) 760000
(c) 1000001   (d) −0.01003

Solution:
(a) $0.0000567 = 5.67/100000 = 5.67 \times 10^{-5}$

(b) $760000 = 7.6 \times 100000 = 7.6 \times 10^{5}$

(c) $1000001 = 1.000001 \times 1000000 = 1.000001 \times 10^{6}$

(d) $-0.01003 = -1.003/100 = -1.003 \times 10^{-2}$

Exercises

3.7 Write each of the following numbers in scientific notation:
    (a) 0.000332   (b) 1040000
    (c) 200030   (d) -0.006004

Rounding

When a decimal number is truncated to either a pre-stated number of decimal places or significant figures the result is an approximation to the original number. A third method of finding approximations to a given number is referred to as **rounding** where a given decimal number is written to the nearest stated power of ten. For example, the number 1546.8 rounded to the nearest whole number is:

1547

and rounded to the nearest 100 is:

1500

When rounding a number it is not necessarily truncated but a similar rule applies to that of defining the last significant figure.

Worked Examples

3.8 Round each of the following decimal numbers to the nearest tenth, unit and thousand:
    (a) 94465.334   (b) -578.36
    (c) 1444.45   (d) -499.99

Solution:
(a) 94465.334: 94465.300, 94465.000, 94000.000

(b) -578.36: -578.40, -578.00, -1000.00

(c) 1444.45: 1444.50, 1444.00, 1000.00

(d) -499.99: -500.00, -500.00, $-0.00 = 0.00$

Exercises

3.8 Round each of the following decimal numbers to the nearest tenth, unit and
thousand:
(a) 10123.549   (b) −638.491
(c) 1499.99   (d) −500.01

Uniqueness
A further feature of the decimal representation is that not all real numbers have a
unique decimal representation. For example, the number:

$$0.\dot{9} = 0.9999....$$

when truncated either to any number of significant figures or to any number of
decimal places results in the number 1. As a consequence, either:

$$1.\dot{0}$$
or
$$0.\dot{9}$$

can be taken to represent the number 1.

Worked Examples

3.9 Find two different decimal representations of each of the following numbers:
(a) 1/5   (b) −3/4
(c) 1/20   (d) −7/32

Solution:

(a) 1/5: 0.2 and 0.1$\dot{9}$

(b) −3/4: −0.75 and −0.74$\dot{9}$

(c) 1/20: 0.05 and 0.04$\dot{9}$

(d) −7/32: −0.21875 and −0.21874$\dot{9}$

Exercises

3.9 Find two different decimal representations of each of the following numbers:
(a) 3/10   (b) −1/4
(c) 3/25   (d) −5/16

Unit 3   The arithmetic of decimals

Try the following test:

1   Evaluate each of the following by longhand. Check your answers using a calculator:
    (a)   28.5381 + 391.477   (b)   85.0604 + 93.949 + 10.0063
    (c)   632.794 – 524.895   (d) 27.41×45.32
    (e)   659.4348 ÷ 43.27

2   Evaluate each of the following using a calculator:
    (a)   32.5464 + 72.59421 – 18.605   (b) $256.42 \times 18.36 \times 11.1243$
    (c) $84.61 \times 0.094 \div 12.61$ (d) $(105.501 \div 32.63) \div 18.504$
    (e) $11^{0.11}$ (f) $23.25^{-1.23}$

3   Convert each of the following into a decimal number:
    (a)   2/9   (b)   –15/43
    (c)   19/12   (d)   –17/24

4   Convert each of the following into a fraction in its lowest terms:
    (a)   0.512   (b)   –6.48
    (c)   0.012   (d)   –18.45

The arithmetic of decimal numbers

We have previously seen how to combine rational numbers under the arithmetic operations. In what follows we shall describe how decimal numbers are combined under the arithmetic operations using either longhand methods or an electronic hand calculator. When handling numbers in longhand it is necessary to be aware of place value order and to write the numbers down in a very specific format.

Addition

The process of the longhand addition of two numbers of any size requires only the ability to add any pair of single numeral numbers together coupled with the notion of **carrying over** to the next column. For example, to perform the addition 9 + 3 we write the second number immediately beneath the first thus:

$$
\begin{array}{r}
1 \\
9 \\
+ \; 3 \\
\hline
12
\end{array}
$$

Here the sum of 9 and 3 is 12 where the 2 is written in the units column and the 10 is represented by **carrying over** the 1 to the tens column. This can be represented by a 1 to the top left as indicated. This principle can be employed repetitively for the addition of any two numbers. For example the addition:

53.67 + 1.547

is performed by writing the numbers one beneath the other with their decimal points aligned. If desired, additional zeros can be inserted at either end of each number to enable the addition to be performed in columns as follows:

$$\begin{array}{r} \scriptstyle 1\ \ 1 \\ 53.670 \\ +\ 01.547 \\ \hline 55.217 \end{array}$$

With practice it soon becomes possible to add more than two numbers together in one sum:

$$\begin{array}{r} \scriptstyle 1\,2\ \ 2\,1 \\ 24.789 \\ 267.770 \\ 1.556 \\ \hline 294.115 \end{array}$$

Subtraction
The subtraction of two numbers is performed in a manner quite analogous to that of addition with the substitution of carrying by **taking**. For example the subtraction 35 – 18 is performed as follows:

Write the numbers beneath each other so that units are beneath units and tens beneath tens thus:

$$\begin{array}{r} 35 \\ -18 \end{array}$$

The subtraction then performed column by column. However, we cannot take 8 from 5 without invoking negative numbers. Consequently, we **take** a ten from the tens column and add it to the five to make 15 thereby reducing the number in the tens column from 3 to 2. The subtraction can then be effected thus:

$$\begin{array}{ccc} 35 & & \overset{\scriptstyle 1}{2}5 \\ -18 & \rightarrow & -18 \\ & & \hline \\ & & 17 \end{array}$$

This procedure can be employed repetitively for the subtraction of any two numbers. For example the subtraction 50.61 – 1.547 is performed progressively as follows:

		1	11	11	1 11
50.61	50.610	50.600	50.500	50.500	40.500
−1.547	− 01.547	- 01.547	- 01.547	- 01.547	- 01.547
		3	63	.063	49.063

Notice that in the first rewrite of this subtraction the two numbers are arranged so that their respective decimal points are beneath each other.

Multiplication
Multiplication was defined as repetitive addition and this is the procedure employed when multiplying two decimal numbers. For example, the product:

$$12.45 \times 31.6$$

is evaluated as follows:

```
    12.45
  ×  31.6
  373.500        12.45 × 30
   12.450        12.45 × 1
    7.470        12.45 × 0.6
  393.420        12.45 × (30 + 1 + 0.6)
```

That is:

$$12.45 \times 31.6 = 393.42$$

Notice that in the body of the working all the numbers found by multiplication are written to 3 decimal places – *the sum of the decimal places of the two numbers involved in the product.* This ensures that the decimal points will all be beneath each other thereby making the final addition easier to handle.

Division
Division was defined as repetitive subtraction and this is the procedure employed when dividing two decimal numbers. For example the division:

$$393.42 \div 31.6$$

is evaluated as follows:

```
            12.450        the number of subtractions are recorded on this top line
  31.6 |⎺  393.420
       − 316.000          − 10 × 31.6
          77.420
        − 63.200          − 2 × 31.6
          14.220
        − 12.640          − 0.4 × 31.6
```

$$1.580$$
$$-\ 1.580 \qquad\qquad -\ 0.05 \times 31.6$$
$$\overline{0} \qquad\qquad -\ 12.45 \times 31.6$$

Because 31.6 can be taken away 12.45 times from 393.42 leaving a zero remainder we see that:

$$393.42 = 31.6 \times 12.45$$

so that:

$$393.42 \div 31.6 = 12.45$$

The decimal form of a number can be converted into the fractional form by expressing each decimal coefficient as a fraction:

$$3.125$$

can be written as the sum:

$$3 \times 10^{0} + 1 \times 10^{-1} + 2 \times 10^{-2} + 5 \times 10^{-3}$$
$$= 3 + 1/10 + 2/100 + 5/1000$$
$$= 3000/1000 + 100/1000 + 20/1000 + 5/1000$$
$$= 3125/1000$$
$$= 25/8$$

So that 25/8 is the fractional version of the decimal 3.125.
In addition, the fractional form of a number can be converted into the decimal form by using the process of division. For example, the number:

$$3.125$$

is the decimal version of the fraction 25/8 and can be obtained from 25/8 by performing the division:

```
        3.125
8 ⌐ 25.000
      24.000
       1.000
       0.800
       0.200
       0.160
       0.040
       0.040
       0.000
```

By these methods any fraction can be converted to its decimal form and any decimal number can be converted to its fractional form.

Worked Examples

3.10 Evaluate each of the following by longhand. Check your answers using a calculator:
(a) 18.7566 + 254.885   (b) 342.8556 + 34.655 + 8.0054
(c) 734.688 − 255.889   (d) 34.58 × 22.73
(e) 7940.1676 ÷ 119.34

Solution:

(a)
```
        1 1  1 1
      18.7566
    + 254.8850
      273.6416
```

(b)
```
       1 1  1 1 1
     342.8556
      34.6550
    +  8.0054
     385.5160
```

(c)
```
                          1 1  1 1 1
      734.688           623.578
    − 255.889         − 255.889
                        478.799
```

(d)
```
        34.58
      × 22.73
      691.6000
       69.1600
       24.2060
        1.0374
      786.0034
```

(e)
```
                       66.534
        119.34 ⌐ 7940.16760
                 7160.40000
                  779.76756
                  716.04000
                   63.72756
                   59.67000
                    4.05756
                    3.58020
                    0.47736
                    0.47736
                          0
```

Exercises

3.10 Evaluate each of the following by longhand. Check your answers using a calculator:
(a) $24.6673 + 102.522$ (b) $19.0013 + 112.447 + 9.01$
(c) $497.367 - 194.538$ (d) 18.63×39.74
(e) $3754.1248 + 35.67$

Using a calculator
By far the most convenient way to perform the arithmetic of decimal numbers is to use an electronic hand calculator. With such a calculator two numbers are combined using any of the arithmetic operation keys by:

Entering the first number using the number keys
Pressing the operation key
Entering the second number
Pressing the = key.

As these procedures are executed the appropriate number appears in the calculator panel.

Raising to a power
The process of raising a decimal number to a decimal power is the fifth remaining operation between decimal numbers that has yet to be discussed. To perform this operation longhand is not a simple task and for this reason we shall use a hand calculator by using the x^y key. For example, to find the value of:

$$25^{0.5}$$

Enter the number 25
Press the x^y key
Enter the number 0.5
Press the = key
The result, 5, is then displayed

Worked Examples

3.11 Evaluate each of the following using a calculator:
(a) $23.4456 + 74.55634 - 32.48995$ (b) $3775.44 \times 886.564 \times 55.674$
(c) $94.55 \times 66.734 + 55.24$ (d) $(998.673 + 66.755) + 55.5576$
(e) $12^{0.32}$ (f) $18.65^{-0.344}$

Solution:
(a) $23.4456 + 74.55634 - 32.48995$: 65.51199

(b) $3775.44 \times 886.564 \times 55.674$: 1.863503×10^8 to 7 significant figures

(c) $94.55 \times 66.734 + 55.24$: 114.22338 to 5 decimal places

(d) $998.673 \div 66.755 \div 55.5576$: 0.269275 to 6 decimal places

(e) $12^{0.32} = 2.2148178$ to 7 decimal places

(f) $18.65^{-0.344}$ to evaluate this will require the use of the calculator's memory:

Enter 0.344 and press the +/- key to change the display to -0.344
Enter this number into the memory (press **MIN**)
Enter the number 18.65
Press the x^y key
Retrieve the memory – press **MR** – to display -0.344
Press the = key to display the result:

$= 0.3654992$ to 7 decimal places

3.12 Convert each of the following into a decimal number:
    (a) 5/6   (b) –18/32
    (c) 15/4   (d) –22/8

Solution:
(a)

$$
\begin{array}{r}
0.833... \\
6\,\overline{)\,5.00000} \\
4.80000 \\
\hline
0.20000 \\
0.18000 \\
\hline
0.0200 \\
0.0180 \\
\hline
0.0020 \\
\cdots\cdots
\end{array}
$$

(b) $-18/32 = -(18/32) = -(0.5625) = -0.5625$

(c) $15/4 = 3.75$

(d) $-22/8 = -2.75$

3.13 Convert each of the following into a fraction in its lowest terms:
    (a) 3.125   (b) –0.554
    (c) 0.024   (d) –55.35

Solution:
(a) 3.125 = 3125/1000 = 125/40 = 25/8

(b) −0.554 = −(554/1000) = −(277/500) = −277/500

(c) 0.024 = 24/1000 = 3/125

(d) −55.35 = −5535/100 = −1107/50

Exercises

3.11 Evaluate each of the following using a calculator:

(a) 16.5379 + 27.44592 − 0.00324   (b) 1945.23 × 765.82 × 22.6538
(c) 13.67 × 22.754 ÷ 19.52   (d) (627.519 ÷ 55.826) + 122.674
(e) $15^{0.67}$ (f) $38.455^{-0.274}$

3.12 Convert each of the following into a decimal number:
(a) 7/12   (b) −19/31
(c) 17/8   (d) −39/15

3.13 Convert each of the following into a fraction in its lowest terms:
(a) 0.3125   (b) −0.128
(c) 0.036   (d) −31.35

Module 3   Further exercises

1   Write each of the following numbers in the form of sum of coefficients multiplying 10 to the appropriate power:
    (a) 472901   (b) −425
    (c) 99.908   (d) 10.0001

2   By finding equivalent fractions convert each of the following fractions to decimal numbers:
    (a) 7/20   (b) 5/8
    (c) 13/2   (d) −9/5

3   Write each of the following decimal numbers in a shorthand form:
    (a) 5/13 = 0.38461538461538...   (b) 7/11 = 0.636363...
    (c) −5/12 = −0.41666666...   (d) −4/27 = −0.148148148...

4   In each of the following identify the numeral in the 1000th decimal place:
    (a) 5/13 = 0.38461538461538...   (b) 7/11 = 0.636363...
    (c) −5/12 = −0.41666666...   (d) −4/27 = −0.148148148...

5   Truncate each of the following decimal numbers to one, two and three decimal places:
    (a) 12.3456   (b) −0.015468
    (c) 7.11111   (d) −36.0909

6   Truncate each of the following decimal numbers to three, four and five significant figures:
    (a) 56432.105   (b) −501.105
    (c) 2398.89   (d) −999.99

7   Round each of the following decimal numbers to the nearest tenth, unit and thousand:
    (a) 56432.105   (b) −501.105
    (c) 2398.89   (d) −999.99

8   Find two different decimal representations of each of the following numbers:
    (a) 3/5   (b) −5/2
    (c) 7/25   (d) −9/16

9   Evaluate each of the following by longhand. Check your answers using a
     calculator:
     (a) 75.8234 + 98.709   (b) 17.9816 + 213.543 + 10.09
     (c) 794.763 − 491.835   (d) 81.36 × 73.94
     (e) 748.2585 ÷ 67.35

10. Evaluate each of the following using a calculator:
     (a) 65.1397 + 47.24925 − 4.0023   (b) 4951.36 × 578.62 × 62.5238
     (c) 6.137 × 5.2724 ÷ 5.921   (d) (719.516 + 82.655) ÷ 67.4212
     (e) $12^{-0.65}$ (f) $43.852^{-2.407}$

11. Convert each of the following into a decimal number:
     (a) 7/11   (b) −15/31
     (c) 19/8   (d) −17/15

12. Convert each of the following into a fraction in its lowest terms:
     (a) 0.624   (b) −8.63
     (c) 0.072   (d) −17.65

Module 4

Binary and hexadecimal numerals

OBJECTIVES

When you have completed this module you will be able to:

- Convert decimal numbers to their binary and hexadecimal representations

- Manipulate the arithmetic of numbers in binary and hexadecimal representations

---

There are four units in this module:

Unit 1: Binary numbers
Unit 2: Negative binary numbers
Unit 3: Adding and subtracting binary numbers
Unit 4: Base 16

Unit 1 Binary numbers

Try the following test:

1   Write down the decimal form of each of the following binary numbers:
    (a) 101   (b) 1100
    (c) 11011   (d) 101101

2   Convert each of the following decimal numbers into their binary form:
    (a) 9   (b) 13
    (c) 201   (d) 101

3   The following numbers are written in the decimal representation. Find the representation in base 4:
    (a) 7   (b) 12
    (c) 196   (d) 444

4   The following numbers are written in base 4. Find their binary form:
    (a) 11   (b) 23
    (c) 301   (d) 3333

The number base 2

It cannot be repeated too often that there is an important distinction between a number and the numerals used to represent that number. It is not easy fully to comprehend this statement when the only symbols that we have are those with which we are familiar and which we use every day. We continually confuse the numeral *2* with the word *two* and think of them both as being the same thing. They are not. They are each representatives of the same number but *2* is a written numeral and *two* is a verbal numeral. A pretty harsh distinction to make, you may think, but one that is necessary. To an ancient Roman the written numeral was *II* and the verbal numeral was *duo* but the number was still the same number that we have today.

The German philosopher Nietzsche said:

> '*We always express our thoughts with the words that lie to hand. Or, to express my whole suspicion: we have at any moment only the thoughts for which we have to hand the words.*'

We can think of numbers only in words or mental symbolic pictures and the effect is that the numerals we use to describe the idea of number become confused with the idea itself. With time and continual use this fusion becomes seamless, lending an artificial unity to the two concepts of number and numeral.

All the numbers we have discussed have been described using sequences of numerals taken from a list of ten, 0, ..., 9. The reason for using ten numerals is unclear though many would claim it was because we have ten fingers. Whether this is true or not is irrelevant. What is clear is that the choice of ten symbols is quite arbitrary.

Here is a little tale.

Smallville is a mid-west township in Midstate, USA and in the centre of Main Street is a Soda Fountain where the locals go to quench their thirst during the long, hot and dusty days of summer. The Soda Fountain is owned by a man called Wishbone whose young son Joey attends behind the bar. One bright morning in late Spring young Joey's father took delivery of a brand new dispenser – a sparkling, polished, steel and chromium affair with three brass switches labelled A, B and C at the front and immediately beneath the legend *Refreshing Lime 'n Lemon*. Beneath the three brass switches was a tap through which a drink could be dispensed simply by flicking up combinations of switches.

Flick up the left-hand switch and soda would come shooting through the tap. The middle switch produced concentrated lemon juice and the right-hand switch dispensed lime cordial. Joey was entranced by this new machine. He was fascinated with the way it worked and pretty soon he had figured out how many different combinations of drinks he could serve from it. They were:

A	B	C	Drink
down	down	down	Nothing
down	down	up	1 Lime cordial
down	up	down	2 Lemon concentrate
down	up	up	3 Lime and lemon concentrate mix
up	down	down	4 Soda
up	down	up	5 Lime soda
up	up	down	6 Lemonade
up	up	up	7 Lime 'n lemon soda

Joey's problem was remembering which combination of switches produced which drink until he realized there was a pattern. He decided to give each switch a code number:

> when switch A was **up** it had the code number **4**
> when switch B was **up** it had the code number **2**
> when switch C was **up** it had the code number **1**
> when any switch was **down** it had the code number **0**

Using this code he could instantly link drinks to numbers and numbers to switch combinations. For instance, if someone came in and asked for Lemonade he knew that Lemonade was number 6 on the list and that:

$$6 = 4 + 2 + 0 \ or$$

switch A **up**, switch B **up** and switch C **down**.

So successful was this idea that he eventually changed the name of the Lime 'n Lemon soda to **Seven-Up.**

Replacing the ups and downs in the table with the appropriate numbers we find the sense behind Joey's idea.

A	B	C		Drink
0	0	0	0	Nothing
0	0	1	1	Lime cordial
0	2	0	2	Lemon concentrate
0	2	1	3	Lime and lemon concentrate mix
4	0	0	4	Soda
4	0	1	5	Lime soda
4	2	0	6	Lemonade
4	2	1	7	Lime and lemon soda

Simple addition of switch numbers produces the drink number. Looking a little closer we see that each column contains only two different numbers, one of which is zero. Recognizing that:

$$4 = 2^2, 2 = 2^1 \text{ and } 1 = 2^0$$

we can rewrite these switch combinations as follows:

$$
\begin{array}{ccc}
\mathbf{A} & \mathbf{B} & \mathbf{C}
\end{array}
$$
$$0 \times 2^2 + 0 \times 2^1 + 0 \times 2^0 = 0$$
$$0 \times 2^2 + 0 \times 2^1 + 1 \times 2^0 = 1$$
$$0 \times 2^2 + 1 \times 2^1 + 0 \times 2^0 = 2$$
$$0 \times 2^2 + 1 \times 2^1 + 1 \times 2^0 = 3$$
$$1 \times 2^2 + 0 \times 2^1 + 0 \times 2^0 = 4$$
$$1 \times 2^2 + 0 \times 2^1 + 1 \times 2^0 = 5$$
$$1 \times 2^2 + 1 \times 2^1 + 0 \times 2^0 = 6$$
$$1 \times 2^2 + 1 \times 2^1 + 1 \times 2^0 = 7$$

This representation is entirely analogous to the decimal representation where now, instead of having coefficients multiplying powers of 10 we have coefficients multiplying powers of 2. Just as we suppressed the powers of 10 to form the decimal representation of a number so we can suppress the powers of 2 and write the representation of our numbers as:

$000 = 0$		$100 = 4$
$001 = 1$		$101 = 5$
$010 = 2$		$110 = 6$
$011 = 3$		$111 = 7$

The representation on the left, being based upon the number 2, is called the **binary representation** of the number system. Again, the numerals 0 and 1 are located on place value order where each place, or column, is assumed to be headed by 2 raised to the appropriate power.

Notice that in this last column of equalities we are comparing different symbolic representations of the same numbers. To make this explicitly clear we should introduce the **suffix** 2 in the left-hand numbers and the suffix 10 in the right hand numbers in order to emphasize that we are using different number bases:

$$000_2 = 0_{10}$$
$$001_2 = 1_{10}$$
$$010_2 = 2_{10}$$
$$011_2 = 3_{10}$$
$$100_2 = 4_{10}$$
$$101_2 = 5_{10}$$
$$110_2 = 6_{10}$$
$$111_2 = 7_{10}$$

This sort of notation tends to become cumbersome after a while and the suffix is normally omitted when the value of the base used is clear.

Binary numbers

A binary number is a sequence of 0's and 1's where each 0 and 1 is the coefficient of 2 to some power. For example, the binary number:

10111

represents the number:

$$1 \times 2^4 + 0 \times 2^3 + 1 \times 2^2 + 1 \times 2^1 + 1 \times 2^0$$

By recognizing that:

$$2^4 = 16$$
$$2^3 = 8$$
$$2^2 = 4$$
$$2^1 = 2$$
$$2^0 = 1$$

the binary representation 10111 can be converted to its decimal representation as follows:

$$\begin{aligned}
1 \times 2^4 + 0 &\times 2^3 + 1 \times 2^2 + 1 \times 2^1 + 1 \times 2^0 \\
&= 1 \times 16 + 0 \times 8 + 1 \times 4 + 1 \times 2 + 1 \times 1 \\
&= 16 + 0 + 4 + 2 + 1 \\
&= 23
\end{aligned}$$

By dividing 23 by 2 successively we can retrieve the binary form:

Remainder

$$
\begin{array}{c|cc}
2 & 23 & \\
2 & 11 & 1 \\
2 & 5 & 1 \\
2 & 2 & 1 \\
2 & 1 & 0 \\
\hline
 & 0 & 1 \\
\end{array}
$$

In other words:

$$
\begin{aligned}
23 &= 2 \times 11 + 1 \\
&= 2 \times (2 \times 5 + 1) + 1 \\
&= 2 \times (2 \times (2 \times 2 + 1) + 1) + 1 \\
&= 2 \times (2 \times (2 \times (2 \times 1 + 0) + 1) + 1) + 1 \\
&= 1 \times 2^4 + 0 \times 2^3 + 1 \times 2^2 + 1 \times 2^1 + 1 \times 2^0
\end{aligned}
$$

By taking the remainders in the **reverse order** in which they appear as the division progresses we retrieve the binary form of the number whose decimal form is 23.

Worked Examples

4.1 Write down the decimal form of each of the following binary numbers:

(a) 101   (b) 1101
(c) 10001   (d) 101010

Solution:

(a) Expanding the binary number 101 as a sum of powers we find that:

$$
\begin{aligned}
101 &= 1 \times 2^2 + 1 \times 2^1 + 1 \times 2^0 \\
&= 4 + 2 + 1 \\
&= 7 \text{ the decimal form.}
\end{aligned}
$$

(b) Expanding the binary number 1101 as a sum of powers we find that:

$$
\begin{aligned}
1101 &= 1 \times 2^3 + 1 \times 2^2 + 0 \times 2^1 + 1 \times 2^0 \\
&= 8 + 4 + 0 + 1 \\
&= 13 \text{ the decimal form.}
\end{aligned}
$$

(c)
$$
\begin{aligned}
10001 &= 1 \times 2^4 + 0 \times 2^3 + 0 \times 2^2 + 0 \times 2^1 + 1 \times 2^0 \\
&= 16 + 0 + 0 + 0 + 1 \\
&= 17 \text{ the decimal form.}
\end{aligned}
$$

(d)
$$
\begin{aligned}
101010 &= 1 \times 2^5 + 0 \times 2^4 + 1 \times 2^3 + 0 \times 2^2 + 1 \times 2^1 + 01 \times 2^0 \\
&= 32 + 0 + 8 + 0 + 2 + 0 \\
&= 42 \text{ the decimal form.}
\end{aligned}
$$

4.2 Convert each of the following decimal numbers into their binary form:
  (a) 6   (b) 12
  (c) 243   (d) 100

Solution:
(a) The decimal form 6 can be written as:

$$6 = 4 + 2$$
$$= 1 \times 2^2 + 1 \times 2^1 + 0 \times 2^0$$

This gives the binary form as 110. Alternatively, we can divide successively by 2 and form the binary representation from the remainders:

```
2 | 6
2 | 3                    remainder 0
2 | 1                    remainder 1
    0                    remainder 1
```

In reverse order the remainders are 110 which is the binary form of the decimal number 6

(b) Dividing successively by 2:

```
2 | 12
2 | 6                    remainder 0
2 | 3                    remainder 0
2 | 1                    remainder 1
    0                    remainder 1
```

In reverse order the remainders are 1100 which is the binary form of the decimal number 12

(c)
```
2 | 243
2 | 121                  remainder 1
2 | 60                   remainder 1
2 | 30                   remainder 0
2 | 15                   remainder 0
2 | 7                    remainder 1
2 | 3                    remainder 1
2 | 1                    remainder 1
    0                    remainder 1
```

In reverse order the remainders are 11110011 which is the binary form of the decimal number 243

(d)
```
2 | 100
2 | 50                   remainder 0
```

2	25		remainder 0
2	12		remainder 1
2	6		remainder 0
2	3		remainder 0
2	1		remainder 1
	0		remainder 1

1100100 is the binary form of the decimal number 100

4.3 The following numbers are written in the decimal representation. Find the representation in base 4:
(a) 8   (b) 15
(c) 345   (d) 1291

Solution:

(a) Numbers are represented in base 4 by using the four numerals 0, 1, 2 and 3 as coefficients multiplying respective powers of 4. The number 8 can be written as:

$$2 \times 4 = 2 \times 4^1 + 0 \times 4^0$$

This gives the base 4 representation as:

20

(b) The number 15 can be written as:

$$15 \ = 12 + 3$$
$$= 3 \times 4^1 + 3 \times 4^0$$

This gives the base 4 representation as 33.

(c) To convert 345 decimal to base 4 we can divide successively by 4 in a similar manner to the process employed when converting decimal numbers to binary form:

4	345		
4	86		remainder 1
4	21		remainder 2
4	5		remainder 1
4	1		remainder 1
	0		remainder 1

Therefore, listing the remainders in reverse order:

11121 is the base 4 form of the decimal number 345. As a check, notice that:

$$11121 = 1 \times 4^4 + 1 \times 4^3 + 1 \times 4^2 + 2 \times 4^1 + 1 \times 4^0$$
$$= 256 + 64 + 16 + 8 + 1$$
$$= 345$$

(d) 1291 decimal converts to base 4 form as follows:

4	1291	
4	322	remainder 3
4	80	remainder 2
4	20	remainder 0
4	5	remainder 0
4	1	remainder 1
	0	remainder 1

Therefore, listing the remainders in reverse order:

110023 is the base 4 form of the decimal number 345.

As a check, notice that:

$$110023 = 1 \times 4^5 + 1 \times 4^4 + 0 \times 4^3 + 0 \times 4^2 + 2 \times 4^1 + 3 \times 4^0$$
$$= 1024 + 256 + 8 + 3$$
$$= 1291$$

4.4 The following numbers are written in base 4. Find their binary form:
(a) 3   (b) 12
(c) 322   (d) 1231

Solution:
(a) The numeral 3 represents the same number in the base 4 representation as it does in the decimal representation. The binary form is, therefore:

11

(b) The number represented by 12 in base 4 is represented as:

$1 \times 4^1 + 2 \times 4^0 = 6$ in decimal and the binary form of decimal 6 is 110.

(c) Each digit of a base 4 representation can be represented in binary form as a two digit numeral:

$0 = 00$
$1 = 01$
$2 = 10$
$3 = 11$

To convert a base 4 representation of a number into its binary representation simply convert each digit in the base 4 form into its binary equivalent:

This means that:

322 becomes 111010 which is the binary representation of the number 322 in base 4 representation. The following will explain why this is so:

$$322 = 3 \times 4^2 + 2 \times 4^1 + 2 \times 4^0$$
$$= 3 \times 2^4 + 2 \times 2^2 + 2 \times 2^0$$

Because $4 = 2^2$ this sequence of powers of 2 only has even powers. Each coefficient can be represented by a 2 digit binary numeral thereby generating the odd powers:

$$= (1 \times 2^1 + 1 \times 2^0) \times 2^4 + (1 \times 2^1 + 0 \times 2^0) \times 2^2 + (1 \times 2^1 + 0 \times 2^0) \times 2^0$$
$$= 1 \times 2^5 + 1 \times 2^4 + 1 \times 2^3 + 0 \times 2^2 + 1 \times 2^1 + 0 \times 2^0$$

(d) Each numeral of 1231 in base 4 representation becomes, in the binary representation:

01, 10, 11, 01 respectively.

This gives the binary representation of 1231 in base 4 as:

01101101

Exercises

4.1 Write down the decimal form of each of the following binary numbers:
   (a) 110   (b) 1111
   (c) 10101   (d) 110011

4.2 Convert each of the following decimal numbers into their binary form:
   (a) 7   (b) 17
   (c) 175   (d) 111

4.3 The following numbers are written in the decimal representation. Find the representation in base 4:
   (a) 9   (b) 17
   (c) 234   (d) 2805

4.4 The following numbers are written in base 4. Find their binary form:
   (a) 2   (b) 21
   (c) 123   (d) 2112

Unit 2   Negative binary numbers

Try the following test:

1   Write each of the following negative numbers as an 8-bit binary
    number in:
    (i)   sign-magnitude form
    (ii) complementary form
    (iii) true complement form

    (a) −7   (b) −34   (c) −105

Binary numbers and computers

Why should we need to write our numbers in binary form when we have managed
so well for so long by using the decimal system? That is just the point – *we* have
managed so well but now computers are available to take all the tedium out of our
arithmetic problems; we no longer need to manage but computers do and a proper
understanding of the working of a computer is not possible without a familiarity with
the binary number system.

A computer's memory consists of arrays of switches in much the same manner
as Joey Wishbone's *Lime 'n Lemon* soda dispenser which had a simple array of three
manually operated switches. Unlike Joey's dispenser the switches in a computer's
memory are controlled electronically and are far more numerous. However, the
principle is the same. When a switch is closed current can flow but when a switch is
open the current cannot flow. Again, just like Joey's dispenser, we can describe the
state of a switch by using the numerals 0 and 1 – the state of an open switch is
represented by 0 and the state of a closed switch is represented by 1. So if, for
example, in a computer's memory there is an array of 8 switches the combined states
of all 8 switches can be represented by a binary number within the range:

   00000000 to 11111111

If the binary representation of the states of this array of switches is taken to represent
a stored number then we can see that in the decimal representation this unit of
memory can store any one of 256 numbers ranging from 0 to 255. Each such unit of
memory is called a **byte** and the 0 or 1 state of each component switch of the byte
is called a **bit** – derived from **binary digit**. The number of bits in the byte restricts the
size of the numbers that can be dealt with in a computer's memory and if we restrict
ourselves to 8 bit bytes we restrict ourselves to numbers in the range 0 to 255. Within
the real world of computing there ways around this but for our purposes we shall
assume that we are dealing only with 8 digit binary numbers at most.

Negative numbers

If we are to use a computer's memory to store numbers we need to be able to store negative numbers. On a piece of paper we can indicate a negative number quite simply by using a dash but this is not possible in a computer since all we have are states of switches. There are three ways of indicating a negative number in a computer.

Sign-magnitude

Using the sign-magnitude convention the sign of a binary number is indicated by the leading bit where 0 signifies a positive number and 1 signifies a negative number. For example,

00000111 represents 7 in decimal

and

10000111 represents –7 in decimal.

Again this restricts the range of numbers we can store. We can store 255 different numbers but now they are in the range:

00000000 to 01111111 for positive numbers (0 to 127)

and

10000000 to 11111111 for negative numbers (0 to –127 decimal)

Notice that we now have two representations for zero, 00000000 and 10000000, thereby reducing the original 256 possibilities to 255.

Complementary numbers

A complementary number is found by counting backwards from the highest number available. For example, the highest 8 digit binary number is:

11111111

Counting backwards from this number we see that the next three numbers are:

11111110, 11111101 and 11111100.

As we are counting backwards we can use these numbers to denote negative numbers:

11111110 = –1
11111101 = –2
11111100 = –3

The simplest way to obtain the complement of a number is to change all the 0's to 1's and all the 1's to 0's. For example, in 8 digit notation:

00000001 = 1
00000010 = 2
00000011 = 3

Changing 0's to 1's and vice versa we obtain the complements:

11111110 = –1
11111101 = –2
11111100 = –3

Again, the leading bit denotes the sign of the number and the numbers range from:

01111111 = 127 to 10000000 = –127

where 00000000 and 11111111 both represent zero.

True complements
A second method of complementing, true complementation, also known as 2's complement, which avoids the duplicate representation of zero provides a third means of representing a negative number. True complementation is achieved by complementing in the usual manner and then adding 1 to the result. For example, given:

00000001 = 1
00000010 = 2
00000011 = 3

Changing 0's to 1's and vice versa, and adding 00000001 to the result we obtain the true complements:

11111110 + 00000001= 11111111 = –1
11111101 + 00000001= 11111110 = –2
11111100 + 00000001= 11111101 = –3

Here we see that the addition of two binary numbers is effected using the following:

Decimal	Binary
0 + 1 = 1	00 + 01 = 01
1 + 1 = 2	01 + 01 = 10
2 + 1 = 3	10 + 01 = 11
3 + 1 = 4	11 + 01 = 100

We shall say more about the addition of binary numbers in the following unit.

Notice that the complement of $0 = 00000000$ is:

$11111111 + 00000001 = 100000000$

which is a 9 digit number and thereby not permitted within our 8 bit representation. In this situation, where the 8 bit representation overflows into 9 bits we merely discard the overflow bit to leave:

$00000000 = 0$

Hence the complement of zero is the same as zero, thereby removing the ambiguity of representation. Also the numbers now range from:

$01111111 = 127$ to $10000000 = -128$

Removing the two forms for zero means that we have gained an extra number and in the following unit this is the form of complementation that we shall use.

Worked Examples

4.5 Write each of the following negative numbers as an 8-bit binary number in:
   (i)   sign-magnitude form
   (ii) complementary form
   (iii) true complement form

   (a) -5 (b) -28 (c) -103

Solution:
(a) -5

   (i) In binary 00000101 represents 5 so 10000101 represents -5 in sign-magnitude form

   (ii) In binary 00000101 represents 5 so interchanging 0's with 1's and vice versa yields 11111010 which represents -5 in complement form

   (iii) In true complement form -5 is represented by the complement form plus 1. That is:

   $11111010 + 00000001 = 11111011$

(b) -28

   (i) In binary 00011100 represents 28 so 10011100 represents -28 in sign-magnitude form

(ii) In binary 00011100 represents 28 so interchanging 0's with 1's and vice versa yields 11100011 which represents –28 in complement form

(iii) In true complement form –28 is represented by the complement form plus 1. That is:

11100011 + 00000001 = 11100100

(c)–103

(i) In binary 00110011 represents 103 so 10110011 represents –103 in sign-magnitude form

(ii) In binary 00110011 represents 103 so interchanging 0's with 1's and vice versa yields 11001100 which represents –103 in complement form

(iii) In true complement form –103 is represented by the complement form plus 1. That is:

11001100 + 00000001 = 11001101

Exercises

4.5 Write each of the following negative numbers as an 8-bit binary number in:
  (i)   sign-magnitude form
  (ii) complementary form
  (iii) true complement form

  (a) –9   (b) –43   (c) –112

Unit 3   Addition and subtraction

Try the following test:

1   Find the true complement of each of the following binary numbers:
    (a) 10101010   (b) 11100111
    (c) 01101101   (d) 00110101

2   Represent each of the following decimal numbers in 8-bit binary form
    using true complementation for negative numbers:
    (a) 86   (b) −37
    (c) 95   (d) −121

3   Find the decimal equivalent of each of the following 8-bit binary
    numbers where true complementation has been used to represent
    negative numbers:
    (a) 00011101   (b) 10001110
    (c) 10101011   (d) 00101011

4   Find the sum of each of the following binary numbers where true
    complementation has been used to represent negative numbers:
    (a) 01010111, 01110101   (b) 11011011, 01011011
    (c) 11010011, 11011111   (d) 00010101, 11110001

5   Convert each of the following sums in the decimal representation into
    binary form. Complete the sum and convert the result back to decimal
    form:
    (a) 42 + 73   (b) 36 − 49
    (c) −17 + (−12)   (d) 98 − 89

The arithmetic of the binary representation

We are already familiar with the arithmetic of numbers expressed in the decimal
notation where numerals are located using a place value based on powers of 10. The
identical arithmetic of addition, subtraction, multiplication and division is available
within the binary representation where now numerals are located using a place value
based on powers of 2. Note that from now on negative numbers will be represented
using true complementation.

Addition

The process of the addition of two binary numbers is based upon the following
table:

$0 + 0 = 00$
$0 + 1 = 01$
$1 + 1 = 10$

Here we see the notion of carrying to the next column is used whenever the sum of the numbers in one column exceeds 1. The principle embodied within the simple table above can be extended to enable the addition of binary numbers containing an arbitrary number of numerals. For example:

$1101 + 101001$

is best written as:

```
    1   1
   001101
 + 101001
   111110
```

Subtraction

Subtraction, as we know, is the reverse process to addition and just as we can subtract decimal numbers by reversing the process of addition so we can subtract binary numbers in the same way. However, an alternative method of subtracting one positive number from another is to consider the process as one of adding a negative number to a positive number.

For example, in the decimal representation:

$13 - 6 = 13 + (-6) = 7$

In the binary representation this calls for the use of the true complement.

$13 = 00001101$ and $6 = 00000110$

so that:

$-6 = 11111001 + 00000001 = 11111010$

and hence:

```
   00001101
   11111010
  100000111
```

Discarding the leading 1, the overflow bit, gives:

$00000111 = 7$

Within a computer, addition is achieved using an electronic circuit called an **adder**.

Rather than using a different type of circuit to perform subtraction it is clearly more convenient to use true complementation in conjunction with an adder. The same reasoning applies to multiplication and division as both of these operations can be defined in terms of addition.

Worked Examples

4.6 Find the true complement of each of the following binary numbers:
    (a) 01010101   (b) 11001100
    (c) 11011010   (d) 00011110

Solution:
(a) The true complement of 01010101 is found by reversing the 0's and 1's and adding 1. This gives:

    $10101010 + 00000001 = 10101011$

(b) The true complement of 11001100 is:

    $00110011 + 00000001 = 00110100$

(c) The true complement of 11011010 is:

    $00100101 + 00000001 = 00100110$

(d) The true complement of 00011110 is:

    $11100001 + 00000001 = 11100010$

4.7 Represent each of the following decimal numbers in 8-bit binary form using true complementation for negative numbers:
    (a) 73   (b) −95
    (c) 119   (d) −123

Each of these decimal numbers can be converted to binary form by successively dividing by 2 and listing the remainders in reverse order. Alternatively, each number can be written as a sum of powers of 2 and this is what we do here:

Solution:
(a) $73 = 64 + 8 + 1$
        $= 1 \times 2^6 + 1 \times 2^3 + 1 \times 2^0$

    and so:

      64 is 01000000 in binary form
       8 is 00001000 in binary form and
       1 is 00000001 in binary form.

By addition we see that:

73 is 01001001 in binary form.

(b) +95 = 64 + 16 + 8 + 4 + 2 + 1 and:

64 is	01000000 in binary form
16 is	00010000 in binary form
8 is	00001000 in binary form
4 is	00000100 in binary form
2 is	00000010 in binary form
1 is	00000001 in binary form.
95 is	01011111 in binary form.

The true complement of 01011111 is given as:

10100000 + 00000001 = 10100001 which is the binary representation of –95 using true complementation.

(c) 119 = 64 + 32 + 16 + 4 + 2 + 1.

This means that the binary form is 01110111

(d) +123 = 64 + 32 + 16 + 8 + 2 + 1

This means that the binary form is 01111011. The true complement of this binary number is:

10000100 + 00000001 = 10000101 which is the binary representation of –123 using true complementation.

4.8 Find the decimal equivalent of each of the following 8-bit binary numbers where true complementation has been used to represent negative numbers:
(a) 01110001   (b) 10011100
(c) 11001110   (d) 00010110

Solution:
(a) 01110001 = 64 + 32 + 16 + 1 = 113

(b) To identify a negative binary number written in true complemented form we make use of the familiar idea in the decimal representation that the negative of a negative number is the positive counterpart. For example:

–(–9) = 9

The true complement of a negative, true complemented binary number is the positive equivalent of the binary number.

10011100 is a negative number using true complementation. The positive counterpart is found by finding the true complement of this complement:

The true complement of 10011100 is given as:

01100011 + 00000001 = 01100100

The decimal form of 01100100 is 64 + 32 + 4 = 100. Consequently, the binary number:

10011100

is the binary form of the decimal number:

−100

using true complementation.

(c) 11001110 represents a negative number. The complement is:

00110001 + 00000001 = 00110010 which in decimal form is 32 + 16 + 2 = 50. Consequently:

11001110

is the binary form of the decimal number:

−50

using true complementation.

(d) 00010110 represents the number 16 + 4 + 2 = 22 in the decimal representation.

4.9 Find the sum of each of the following binary numbers where true complementation has been used to represent negative numbers:
(a) 01010001, 00010101   (b) 10011100, 01110110
(c) 11001110, 11011011   (d) 01110110, 01101111

Solution:
(a)   01010001
        + 00010101
        ‾‾‾‾‾‾‾‾‾‾
          01100110

(b)   10011100
        + 01110110
        ‾‾‾‾‾‾‾‾‾‾
         100010010

Discarding the leading 9th bit yields the result 00010010

(c)   11001110
     + 11011011
      110101001

Discarding the leading 9th bit yields the result:

10101001

(d)   01110110
     + 01101111
      11100101

Here the sum of two positive numbers yields a negative number, so clearly something is wrong somewhere. The decimal forms of each of the two numbers in the sum are:

118 and 111 respectively.

The sum of these two numbers is 229 which is outside the range of representation using 8-bit binary numbers and true complementation to represent negative numbers. This is called **overflow.**

4.10 Convert each of the following sums in the decimal representation into binary form. Complete the sum and convert the result back to decimal form:
(a) 39 + 47   (b) 52 – 105
(c) –11 + (–61)   (d) 127 – 128

Solution:
(a) 39 + 47 = 86 in decimal. Now:

39 + 47 = (32 + 4 + 2 + 1) + (32 + 8 + 4 + 2 + 1)

In binary this becomes:

00100111 + 00101111 = 01010110

In decimal this number is:

64 + 16 + 4 + 2 = 86 which is in agreement with the earlier result.

(b) 52 – 105 = 52 + (–105) = –53

52 = (32 + 16 + 4) is   00110100 in binary
105 = (64 + 32 + 8 + 1) is   01101001 in binary

so that:

−105 is 10010110 + 00000001 = 10010111 in true complemented binary.

Therefore, adding these two binary numbers together yields:

```
  00110100
+ 10010111
  11001011
```

11001011 is a negative number. To find the positive counterpart of this negative number we must find its true complement which is given as:

00110100 + 00000001 = 00110101

The decimal form of this binary number is:

32 + 16 + 4 + 1 = 53

Therefore the result of the original sum, namely 11001011, has the decimal form of −53 in agreement with the earlier result.

(c) −11 + (−61) = −72

    11 in binary form is   00001011. This means that:

    −11 in binary form is   11110100 + 00000001 = 11110101.

    61 in binary form is   00111101. This means that:

    −61 in binary form is   11000010 + 00000001 = 11000011.

Adding:

```
  11110101
+ 11000011
 110111000
```

Discarding the leading 9th bit yields:

10111000

The true complement of this number is:

01000111 + 00000001 = 01001000

This number has the decimal form:

64 + 8 = 72

This means that the result of the original sum was –72 in agreement with the earlier result.

(d) 127 – 128 = –1

127 in binary form is   01111111
–128 in binary form is   10000000

Their sum is:

11111111 which is a negative number in true complemented form. The true complement of this binary number is:

00000000 + 00000001 = 00000001

which is 1 in the decimal representation. Consequently, the binary number 11111111 has the decimal form –1 in true complementation.

Exercises

4.6 Find the true complement of each of the following binary numbers:
   (a) 01111110   (b) 10000001
   (c) 10110110   (d) 00100110

4.7 Represent each of the following decimal numbers in 8-bit binary form using true complementation for negative numbers:
   (a) 54   (b) –101
   (c) 124   (d) –77

4.8 Find the decimal equivalent of each of the following 8-bit binary numbers where true complementation has been used to represent negative numbers:
   (a) 00101011   (b) 11100101
   (c) 10110100   (d) 00110001

4.9 Find the sum of each of the following binary numbers where true complementation has been used to represent negative numbers:
   (a) 00101101, 01100101   (b) 11100111, 01001010
   (c) 10110010, 10101101   (d) 00001111, 11011011

4.10 Convert each of the following sums in the decimal representation into binary form. Complete the sum and convert the result back to decimal form:
   (a) 51 + 69   (b) 111 – 99
   (c) –38 + (–57)   (d) 128 – 127

Unit 4   Base 16

Try the following test:

1   Convert each of the following decimal numerals into the equivalent
    hexadecimal numeral:
    (a) 17   (b) 87
    (c) 349   (d) 27832104

2   Convert each of the following hexadecimal numerals into the equivalent
    decimal numeral:
    (a) 4C   (b) FB
    (c) D82   (d) FAED

3   Convert each of the following positive binary numerals into the
    equivalent hexadecimal numeral:
    (a) 11101110   (b) 101010101111
    (c) 1011101101   (d) 11100110101100110

4   Convert each of the following hexadecimal numerals into the equivalent
    binary numeral:
    (a) D   (b) 3B5
    (c) 8CE7   (d) 2ABC1

Hexadecimal numbers

In many respects the behaviour of a computer resembles the action of the human brain. Both receive and store information and both retrieve and manipulate stored information. The information that is stored in the brain is said to be stored in our memory just as the information stored by a computer said to be stored in the computer's memory. Whether the manufactured construct and its workings that we understand as a computer bear any resemblance to the evolved construct and workings of that organ called the brain is an unanswered question. What is clear, however, is that in both cases, information stored has to be retrievable otherwise it is of little value. The volatile information stored by a computer consists of the collective electronic states of its memory components. Each one of these memory components, referred to as a byte, has a unique address to enable the computer to access the information that it contains. This address is in the form of a number ranging from 1 to the total number of memory bytes available. As the total number of memory bytes can run into the thousands of millions the numbering system is quite crucial.

The computer can deal only with binary numbers but the binary numeral that represents one million in decimal is too large to be sensibly used by a human being as a reference code. The computer needs to be able to convert the binary address into a numeral system that is more readily accessible to the human operator. Unfortunately, the ten decimal numerals are inappropriate for two reasons. Firstly, there are too few to describe large numbers sensibly and secondly the conversion between binary and decimal is cumbersome for the computer to perform. What is needed is a numeral system with more than the ten elementary numerals of the decimal system and a numeral system that readily converts to and from the binary representation.

The ideal candidate is the hexadecimal numeral system which is a numeral system with base 16 which means that is must have 16 fundamental numerals. These are the ten decimal numerals plus the first six upper case letters of the alphabet:

0, 1, ..., 9, A, B, C, D, E, F

The decimal numeral equivalents of the last six hexadecimal numerals are:

A = 10
B = 11
C = 12
D = 13
E = 14
F = 15

To convert a binary numeral into a hexadecimal numeral the binary number is first split into groups of four digits each and then each four digit group is represented by a single hexadecimal character. For example the decimal numeral 181 converts to the binary numeral 10110101. To find the equivalent hexadecimal numeral we now split this binary numeral into the two groups of four:

1011 and 0101

Now,

1011 = 13 in decimal which is represented by D in hexadecimal

and

0101 = 5 in both decimal and hexadecimal.

The hexadecimal form of this binary number is then:

D5

two digits instead of 8, a much more compact way of signifying the number. Notice that just like the decimal and binary representations, hexadecimal numerals are written down in place value order. For example:

$D5 = 13 \times 16^1 + 5 \times 16^0$

Worked Examples

4.11 Convert each of the following decimal numerals into the equivalent hexadecimal numeral:

(a) 18   (b) 140
(c) 271   (d) 11179533

Solution:
(a) $18 = 1 \times 16^1 + 2 \times 16^0$

The equivalent hexadecimal numeral is then 12

(b) $140 = 8 \times 16^1 + 12 \times 16^0$
$ = 8 \times 16^1 + C \times 16^0$

The equivalent hexadecimal numeral is then 8C

(c) $271 = 1 \times 16^2 + 0 \times 16^1 + 15 \times 16^0$
$ = 1 \times 16^2 + 0 \times 16^1 + F \times 16^0$

The equivalent hexadecimal numeral is then 10F

(d) The equivalent hexadecimal numeral to the decimal numeral 11179533 can be found by successively dividing by 16 and listing the remainders:

```
16 | 11179533
16 |   698720      remainder 13 = D hexadecimal
16 |    43670      remainder 0
16 |     2729      remainder 6
16 |      170      remainder 9
16 |       10      remainder 10 = A hexadecimal
   |        0      remainder 10 = A hexadecimal
```

The equivalent hexadecimal numeral is then:

AA960D

4.12 Convert each of the following hexadecimal numerals into the equivalent decimal numeral:

(a) A1   (b) DE
(c) A4C   (d) FFEA

Solution:
(a) $A1 = A \times 16^1 + 1 \times 16^0$
$ = 10 \times 16^1 + 1 \times 16^0$
$ = 160 + 1$
$ = 161$

(b) $DE = D \times 16^1 + E \times 16^0$
    $= 13 \times 16^1 + 14 \times 16^0$
    $= 208 + 14$
    $= 222$

(c) $A4C = A \times 16^2 + 4 \times 16^1 + C \times 16^0$
    $= 10 \times 256 + 4 \times 16 + 12 \times 1$
    $= 2560 + 64 + 12$
    $= 2636$

(d) $FFEA = F \times 16^3 + F \times 16^2 + E \times 16^1 + A \times 16^0$
    $= 15 \times 4096 + 15 \times 256 + 14 \times 16 + 10 \times 1$
    $= 61440 + 3840 + 224 + 10$
    $= 65514$

4.13 Convert each of the following positive binary numerals into the equivalent hexadecimal numeral:
(a) 11001011   (b) 100111000010
(c) 1101101011   (d) 1011110111011

Solution:
(a) 11001011 = 1100 1011

The decimal equivalent of 1100 is 12 which has the hexadecimal equivalent C
The decimal equivalent of 1011 is 11 which has the hexadecimal equivalent B

the hexadecimal equivalent of 11001011 is therefore:

CB

(b) 100111000010 = 1001 1100 0010

The decimal equivalent of 1001 is 9 which has the hexadecimal equivalent 9
The decimal equivalent of 1100 is 12 which has the hexadecimal equivalent C
The decimal equivalent of 0010 is 2 which has the hexadecimal equivalent 2

the hexadecimal equivalent of 100111000010 is therefore:

9C2

(c) 1101101011 = 0011 0110 1011

Notice the additional two 0's at the beginning of the numeral to ensure that the entire numeral breaks into groups of four digits.

The decimal equivalent of 0011 is 3 which has the hexadecimal equivalent 3

The decimal equivalent of 0110 is 6 which has the hexadecimal equivalent 6
The decimal equivalent of 1011 is 11 which has the hexadecimal equivalent B

the hexadecimal equivalent of 1101101011 is therefore:

36B

(d) 1011110111011 = 0001   0111   1011   1011

Notice the additional three 0's at the beginning of the numeral to ensure that the entire numeral breaks into groups of four digits.

The decimal equivalent of 0001 is 1 which has the hexadecimal equivalent 1
The decimal equivalent of 0111 is 7 which has the hexadecimal equivalent 7
The decimal equivalent of 1011 is 11 which has the hexadecimal equivalent B
The decimal equivalent of 1011 is 11 which has the hexadecimal equivalent B

the hexadecimal equivalent of 1011110111011 is, therefore, 17BB.

4.14 Convert each of the following hexadecimal numerals into the equivalent binary numeral:
   (a) F   (b) A34
   (c) 5D2C   (d) 1AA2F

The last six hexadecimal digits have the following binary equivalents:

   A = 1010
   B = 1011
   C = 1100
   D = 1101
   E = 1110
   F = 1111

Conversion from hexadecimal to binary (and vice versa) is now simply achieved by looking up the appropriate numeral.

Solution:
(a) F converts to 1111 in binary

(b) A34 converts to 1010   0011   0100 = 101000110100

(c) 5D2C converts to 0101   1101   0010   1100 = 0101110100101100

(d) 1AA2F converts to 0001 1010 1010 0010 1111 = 00011010101000101111

Exercises

4.11 Convert each of the following decimal numerals into the equivalent hexadecimal numeral:
(a) 21
(b) 99
(c) 1024
(d) 94203904

4.12 Convert each of the following hexadecimal numerals into the equivalent decimal numeral:
(a) 3D
(b) AF
(c) 69C
(d) EDDA

4.13 Convert each of the following positive binary numerals into the equivalent hexadecimal numeral:
(a) 10010101
(b) 101111001010
(c) 1001110101
(d) 10111101010101110

4.14 Convert each of the following hexadecimal numerals into the equivalent binary numeral:
(a) B
(b) 5C8
(c) 9A0E
(d) FF38A

Module 4   Further exercises

1   Write down the decimal form of each of the following binary numbers:
   (a) 100   (b) 1001
   (c) 10111   (d) 110110

2   Convert each of the following decimal numbers into their binary form:
   (a) 7   (b) 19
   (c) 147   (d) 999

3   The following numbers are written in the decimal representation. Find the
   representation in base 4:
   (a) 6   (b) 13
   (c) 392   (d) 3241

4   The following numbers are written in base 4. Find their binary form:
   (a) 12   (b) 32
   (c) 213   (d) 3113

5   Write each of the following negative numbers as an 8-bit binary number in:
      i) sign-magnitude form
     ii) complementary form
    iii) true complement form

   (a) −4   (b) −33   (c) −95

6   Find the true complement of each of the following binary numbers:
   (a) 00111100   (b) 11000011
   (c) 10011100   (d) 01100110

7   Represent each of the following decimal numbers in 8-bit binary form using
   true complementation for negative numbers:
   (a) 63   (b) −71
   (c) 111   (d) −95

8   Find the decimal equivalent of each of the following 8-bit binary numbers
   where true complementation has been used to represent negative numbers:
   (a) 00011000   (b) 11100111
   (c) 10011001   (d) 01100110

9   Find the sum of each of the following binary numbers where true
    complementation has been used to represent negative numbers:
    (a) 00011000, 11100111   (b) 10011001, 01100110
    (c) 11011011, 10010010   (d) 00001111, 11110000

10 Convert each of the following sums in the decimal representation into binary
    form. Complete the sum and convert the result back to decimal form:
    (a) 21 + 68   (b) 59 – 73
    (c) –18 + (–37) (d) 104 – 105

11 Convert each of the following decimal numerals into the equivalent hexadecimal
    numeral:
    (a) 13   (b) 76
    (c) 949   (d) 11111111

12 Convert each of the following hexadecimal numerals into the equivalent decimal
    numeral:
    (a) B9   (b) BF
    (c) F3A   (d) ABBA

13 Convert each of the following positive binary numerals into the equivalent
    hexadecimal numeral:
    (a) 11001100   (b) 111000111000
    (c) 1101011010   (d) 10011011100011001

14 Convert each of the following hexadecimal numerals into the equivalent binary
    numeral:
    (a) A   (b) 2DF
    (c) A61F   (d) FADED

Part Two

Algebra

At the forefront of the social and cultural development of the human race has been an overwhelming curiosity – a desire to know and a desire to understand. Where this desire springs from is unknown but it is there nonetheless. It above all other attributes accounts for the developments we have made as a species. From the hunter-gathering of *Australopithecus* via the more agrarian existence of *Homo erectus* through to *Homo sapiens* – a species of hominid of which we are all but examples – humankind has evolved to its current state. The ability to reflect backwards in time and to project forward into the future has enabled us to become the most dominant species on this planet. Not least in our armoury of talents is an ability to translate from the concrete to the imaginary and back again; to conjure the past and the future in our imagination – the one major by-product of a curious mind is a fertile imagination. The questions what?, how? and why? very quickly become the conjectures what if?, how can I? and why not? Curiosity coupled with imagination has been the driving force behind the development of mathematical thinking throughout the ages and no better evidence of this exists than the step that was taken from arithmetic to algebra.

In this part the origins of the algebraic method are described alongside the manipulative rules of algebra and many of the ramifications that spring from them.

Module 5

Into algebra

OBJECTIVES

When you have completed this module you will be able to:

■ Solve simple word problems

■ Use alphabetic characters to represent numbers

■ Identify the terms and coefficients in an algebraic expression

■ Recognize the rules of algebra as the rules of arithmetic in general form

■ Eliminate the brackets in an algebraic expression

There are three units in this module:

Unit 1: Word problems
Unit 2: Symbols
Unit 3: Into algebra

Unit 1 Word problems

Try the following test:

1   After staining the holy chaplet of fair-eyed Justice that I might see thee, all-subduing gold, grow so much, I have nothing; for I gave 40 talents under evil auspices to my friends in vain, while, O ye varied mischances of men, I see my enemies in possession of the half, the third, and the eighth of my fortune. (How many talents did the unfortunate man once possess?)

2   When I was going to St Ives
I met a man with seven wives.
Each wife had seven sacks
In each sack were seven cats
And with each cat were seven kits.
Kits, cats, sacks and wives.
How many were going to St Ives?

3   A male bee has only one parent whilst a female bee has two. The family tree of a male bee as far back as great-grandparents is as follows:

(a) Extend the family tree back three generations
(b) List the number of ancestors in each generation
(c) How many ancestors does the male bee have in the next previous four generations?

Now here's a problem ...
I went to a car boot sale where one car was selling teeshirts singly and in bundles of four. A board proclaimed the price of a bundle at £2. At 50p each I reckoned they were a bargain so I selected three single teeshirts - a green one, a red one and a white one. I offered the attendant a five pound note. He took the note and thrust his hand into his pocket, rummaging for change. Eventually, he handed me £2.75.

'Hey' I said. 'You've charged me too much. This board says teeshirts are 50p each and you have charged me 75p each.'

'No.' he replied. 'You only get the discount when you buy four.'

So I picked up a fourth teeshirt and asked him for 25p back.

We are all familiar with this sort of situation where we have to use our arithmetic skills. We average, total and compare numerical quantities, almost without conscious thought. It will be instructive to write down exactly the processes used.

Bundle of teeshirts for £2

By dividing both the bundle and the £2 by 4 it is possible to work out that 1 teeshirt is priced at 50p:

1 bundle costs £2

Dividing everything by 4, it is found that

(1 bundle ÷ 4) costs (£2 ÷ 4) or 1 teeshirt costs 50p

Here use has been made of the facts that:

1 bundle = 4 × (1 teeshirt)

and

£2 = 200p
   = 4 × (50p)

so that:

1 bundle ÷ 4 = 4 × (1 teeshirt) ÷ 4 = 1 teeshirt

and

£2 ÷ 4 = 4 × (50p) ÷ 4 = 50p

You may think that these obvious and rather trite facts have been overstated but they do contain within them the germ of an idea that leads to the world of symbolic algebra.

£2.75 change from £5.00

By immediate subtraction I realize that I have been charged

£5.00 − £2.75 = £2.25

for 3 teeshirts. By division I see that this is

£2.25 ÷ 3 for (3 teeshirts) ÷ 3 or 75p for 1 teeshirt

which is greater than the advertised 50p for 1 teeshirt

Problems like this are met every day where the abstract notion of number is used to effect to argue a case in the world of concrete reality. What if ...

... I want to make a fruit salad but I need some apples and I need some bananas. I go to the supermarket and select two apples and four bananas from the display without checking their individual prices. At the checkout I am charged 100p. The salad was delicious and so the following day I decide to make another. This time, at the supermarket I select four apples and two bananas for which I am charged 140p. Is it possible to find out from this information how much a single apple costs and how much a single banana costs?

Word problems such as this and the previous one have a very long pedigree. Over 4000 years ago the Babylonians were discussing problems of this type, presumably for recreation as stimulating mental exercises. They are called rhetorical problems and their solution would have been found by rhetoric – the art of reasoning by using speech or the written word. A typical solution to this last problem might read as follows:

If, on day two I had bought eight apples and four bananas then I would have been charged 280p. The difference between this purchase and the first day's purchase would have been an extra six apples at an extra charge of 180p. This means that each apple costs 30p. If I now use this information in the first day's purchase I find that the four bananas cost me 100p less the charge for two apples which, at 30p each, is 60p. Consequently, the four bananas cost me 40p which means that bananas cost 10p each.

This method of solving such word problems requires a distinct amount of mental juggling and, to some degree, requires the solver to be able to 'see' the way through the problem before embarking upon the solution. A better method of solving this problem is to attack it methodically by first separating the essential information from the inessential and then **writing down** the essential information in as brief a manner as possible.

Writing down the processes involved in solving a problem such as this was, perhaps, the most important single factor to spur on the early development of symbolic algebra.

A second attempt
The essential information contained within the problem consists of the quantities of fruit and the prices paid:

Day 1: 2 apples and 4 bananas cost 100p
Day 2: 4 apples and 2 bananas cost 140p

Simply by looking at this information I realize that if I had doubled my purchase on the second day I would have ended up with the same number of bananas that I had purchased on the first day and would have doubled the cost. That is:

8 apples and 4 bananas would have cost 280p

If I now compare this information with that of my purchases on the first day then I would find that:

8 apples and 4 bananas cost 280p
2 apples and 4 bananas cost 100p

indicating that the larger cost was due to the larger quantity of apples. That is:

6 apples cost 180p

So that 1 apple costs 30p

Now, if I substitute this information into my first day's purchase I find that since:

2 apples and 4 bananas cost 100p

then, since 2 apples cost 60p, I find that:

4 bananas cost 40p

which tells me that:

1 banana cost 10p

This second approach of writing down the reasoning and the solution has the advantage of eliminating the mental juggling – there is no need to retain a fact in the front of your mind once it is written down. Also, by writing down only the essential information the solution tends to falls out of the written down working rather than being preconceived from the start. The solution does, however, suffer from the aspect of being **over-engineered** in that the essential information is still in its original form of English sentences. Whilst this aspect does not create any difficulties for a simple problem such as this one, in more complicated problems even writing down the essential information in this way may not significantly assist in the derivation of the solution. Here is a problem from the Hindu mathematician *Aryabhata the Elder* who lived during the sixth century AD:

Beautiful maiden with beaming eyes, tell me, as thou understandst the right method of inversion, which is the number multiplied by 3, then increased by three quarters of the product, then divided by 7, diminished by 1/3 of the quotient, multiplied by itself, diminished by 52, by the extraction of the square root, addition of 8, and division by 10 gives the number 2?

Extracting the essential information from this word problem results in the following list of facts:

Which is the number when:

multiplied by 3
then increased by 3/4
of the product then divided by 7

diminished by 1/3 of the quotient
multiplied by itself
diminished by 52
extraction of the square root
addition of 8
division by 10

gives the number 2?

Not that much clearer, is it?

The mental confusion inspired by this and other such problems is the reason why they were, and still are, so popular. Everyone likes a puzzle though some may like them more than others. A number of years ago an English artist called *Kit Williams* wrote a book entitled *Masquerade* in which he laid out a number of clues that, when properly gleaned from the text and the pictures, would lead to a Golden Hare that was buried somewhere in England. The world-wide popularity and almost cult following that the book inspired is testament to the universal popularity of the puzzle.

A third attempt

Even after we have separated the essential from the inessential information in a word problem we still need to reduce the essential information to a more manageable form, if for no other reason than that wax tablets and papyrus are at a premium. The obvious method of doing this is to rid the working of extraneous words by using a shorthand symbolism. We need a few ground rules:

We are required to find the cost of one apple and the cost of one banana. These two costs, therefore, represent two **unknowns** that we are trying to find. We shall use a letter of the alphabet to stand for an unknown cost:

Let the letter A stand for the cost of one apple
Let the letter B stand for the cost of one banana

Notice that we have not included the p that stands for 'pence' – it is not necessary if we understand that we are working in pence at all times.

Finally, we make two more abbreviations:

Let + stand for 'and'
Let = stand for 'costs'

Now, because A represents the cost of one apple we can see that:

$$A + A = 2 \times A$$

stands for the cost of two apples. Similarly:

$$B + B + B + B = 4 \times B$$

stands for the cost of four bananas. We are now ready convert the essential information given in English into symbolic form.

Day 1: $(2 \times A) + (4 \times B) = 100$
Day 2: $(4 \times A) + (2 \times B) = 140$

If we were to double the purchase made on Day 2 we would have:

$$2 \times (4 \times A) + 2 \times (2 \times B) = 2 \times 140$$

That is:

$$(8 \times A) + (4 \times B) = 280$$

Notice that here we have used the fact that:

$$2 \times (4 \times A) = (2 \times 4) \times A = 8 \times A$$

By comparing the hypothetical purchase with the purchase made on Day 1:

$(8 \times A) + (4 \times B) = 280$
$(2 \times A) + (4 \times B) = 100$

we find that the extra apples contribute to the extra cost to the extent that:

$$6 \times A = 180$$

Six apples cost 180p. Therefore, to find the cost of one apple we divide everything by 6:

$$6 \times A \div 6 = 180 \div 6$$

That is:

$A = 30$, the cost of one apple is 30p.

Substituting this value into Day 1's purchases by replacing the letter A with the number 30 we find that:

$$(2 \times A) + (4 \times B) = 100$$

becomes:

$$(2 \times 30) + (4 \times B) = 100$$

That is:

$$60 + (4 \times B) = 100$$

So that, if we take 60 from both sides:

$$60 + (4 \times B) - 60 = 100 - 60$$

we find that:

$$(4 \times B) = 40$$

Four bananas cost 40p. Therefore, dividing everything by 4 we find that:

$$B = 10, \text{ the cost of one banana is 10p.}$$

And so the problem is solved using symbols and the rules of arithmetic explicitly.

But wait a minute, what have we actually done here? To begin with we associated a quantity with a letter of the alphabet and then we converted the English words 'and' and 'costs' into the mathematical symbols '+' and '='. Using letters of the alphabet as symbols to represent numbers is not too radical – after all the numerals themselves are only symbols and we have seen how, over the ages, different numeral symbols have been used to represent the same numbers. The radical step is to convert the English words 'and' and 'costs' into the mathematical symbols '+' and '=' because this is the key which unlocks the problem. It **converts the word problem into a mathematical problem** and in doing so it provides us with the whole set of arithmetic rules which we can use in our search for the solution. Instead of juggling with words and their meanings we are able to manipulate symbols using a well defined collection of procedures.

This method of using symbols to work through a problem to find its solution is fundamental to that branch of mathematics called **algebra**, so named after a book called *Hisab al-jabr walmuqabala*. This book was written by the Persian mathematician *Mohammed ibn Musa al-Khowarizmi* around AD 825 and its contents deal with the study of equations. However, we have come a long way since the ninth century as we shall soon find out when we lay down the formal ground rules for algebraic manipulation in the next and subsequent chapters.

Three examples and exercises follow. Try and work through the examples to obtain a flavour of how such problems are solved but do not become too agitated if you fail to understand why any particular method used was used. Solving problems such as these requires a deal of hindsight brought about largely by experience – an experience that will develop as you progress through the pages of this book.

Worked Examples

5.1 In the following problem write down the essential information in symbolic form and then solve the problem.

On Monday I bought 10000 French Francs and 500 US Dollars for a total of £1650. On Tuesday I bought 5000 French Francs and 750 US Dollars for a total of £1225. How much do French Francs and US Dollars cost in pence?

Solution:
Let

F represent the cost of one French Franc in pence and
D represent the cost of one US Dollar in pence.

The cost in pence for 10000 French Francs is then:

$$10000 \times F$$

and the cost in pence for 500 US Dollars is:

$$500 \times D$$

The total cost was £1650 = 165000p so that:

$$165000 = 10000 \times F + 500 \times D$$

Similarly, from Tuesday's purchase we can say that:

$$122500 = 5000 \times F + 750 \times D$$

The essential information is then summarized in the following:

$$165000 = 10000 \times F + 500 \times D$$
$$122500 = 5000 \times F + 750 \times D$$

To solve the problem we recognize that doubling Tuesday's purchase would be represented by:

$$245000 = 10000 \times F + 1500 \times D$$

What we have done here is to double both sides of the equation that represents Tuesday's purchase. An equation represents a balancing of the left- and right-hand sides so whatever arithmetic operation is applied to one side it must also be applied to the other side to maintain the balance. From this information and Monday's purchase we see that:

$$245000 = 10000 \times F + 1500 \times D$$
$$165000 = 10000 \times F + 500 \times D$$

The greater price is due to the greater number of US Dollars.

$$80000 = 1000 \times D$$

Dividing both sides by 1000 we isolate D on the right-hand side. That is:

$$80000 \div 1000 = 1000 \times D \div 1000$$

Showing that:

$$80 = D$$

The cost of 1 US Dollar is 80p.

Substituting this information into Monday's purchase gives:

$$
\begin{aligned}
165000 &= 10000 \times F + 500 \times D \\
&= 10000 \times F + 500 \times 80 \\
&= 10000 \times F + 40000
\end{aligned}
$$

Subtracting 40000 from both sides we find that:

$$165000 - 40000 = 10000 \times F + 40000 - 40000$$

That is:

$$125000 = 10000 \times F$$

Dividing both sides by 10000 to isolate the F on the right-hand side we see that:

$$
\begin{aligned}
125000 \div 10000 &= 10000 \times F \div 10000 \\
&= F
\end{aligned}
$$

This means that:

$$12.5 = F$$

So that each French Franc costs 12.5p

5.2 How many apples are needed if four persons of six receive one third, one eighth, one fourth and one fifth, respectively, of the total number, while the fifth receives ten apples, and one apple remains left for the sixth person?

Solution:
Let A represent the total number of apples needed. This number of apples is then shared out amongst six people as follows:

Person 1: $A \times 1/3$
Person 2: $A \times 1/8$
Person 3: $A \times 1/4$
Person 4: $A \times 1/5$
Person 5: 10
Person 6: 1

If we add all these portions together we have the total number of apples. That is:

$$(A \times 1/3) + (A \times 1/8) + (A \times 1/4) + (A \times 1/5) + 10 + 1 = A$$

Within the four brackets on the left-hand side we can see that A is a common factor. This means that we can rewrite this information as:

$$A \times (1/3 + 1/8 + 1/4 + 1/5) + 11 = A$$

Now, since:

$$1/3 + 1/8 + 1/4 + 1/5 = 40/120 + 15/120 + 30/120 + 24/120$$
$$= 109/120$$

(where 120 is the LCM of 3, 8, 4 and 5) we see that:

$$A \times (109/120) + 11 = A$$

We now wish to subtract $A \times (109/120)$ from both sides so as to leave the 11 isolated on the left. That is:

$$A \times (109/120) + 11 - A \times (109/120) = A - A \times (109/120)$$

That is:

$$11 = A - A \times (109/120)$$

Now, because $A = A \times 1 = A \times (120/120)$ we can rewrite this as:

$$11 = A \times (120/120) - A \times (109/120)$$
$$= A \times (120/120 - 109/120)$$
$$= A \times (11/120)$$

Multiplying both sides by 120/11 so as to isolate the A on the right-hand side we find that:

$$11 \times (120/11) = A \times (11/120) \times (120/11)$$
$$= A$$

because $(11/120) \times (120/11) = 1$. So that the total quantity A of apples is given as:

$$11 \times (120/11) = A$$

which means:

$$120 = A$$

5.3 A farmer died leaving his estate to his three sons. The estate was to be divided amongst his children in the proportions one half to his eldest son and one sixth to each of the other two. Part of the estate consisted of five horses and the sons were unsure of how to divide them amongst themselves. A neighbour heard of the problem and arrived one day at the paddock with one of his own horses. How was the problem resolved?

Solution:
The neighbour put his horse in the paddock with the other five to make six horses in total. He then gave the eldest son half of the horses – three and to the other two he gave each a sixth – one horse each. He had thus distributed five horses leaving his own behind which he took home with him. Each son was happy because they each had more than the original possibility and the neighbour was happy because he no longer lived next to quarrelsome neighbours.

Exercises

5.1 In the following problem write down the essential information in symbolic form. If you feel confident enough you may attempt to solve the problem.

> To travel from my home to my uncle's house I can start my journey by car and complete it by rail. An alternative route involves a shorter car journey but two rail journeys. If I take the first route then I use three gallons of petrol and the total cost of the journey is £27.50. The alternative route only uses one gallon of petrol and has a total cost of £22.50. If each of the two alternative rail journeys cost one half of the single rail journey how much is petrol per gallon and how much do the rail journeys cost?

5.2 You are presented with two glasses, the first containing 100 ml of water and the second 100 ml of wine. A 5 ml spoonful of water is taken from the first glass and poured into the second glass. The wine–water mixture is then thoroughly stirred so as to completely mix the wine and the water. Next, a 5 ml spoonful of the water-wine mix is taken from the second glass and poured into the water in the first glass. Is the amount of wine in the water in the first glass more or less than the amount of water in the wine glass?

5.3 On January 1st a male rabbit and a female rabbit were born. During the first month of their life they are too immature to mate. These two rabbits mate on February 1st and on March 1st they give birth to another male–female pair. If the rabbits remain monogamous and after the first month of their life breed with their siblings to produce a further breeding pair how many pairs of rabbits are there at the beginning of each month up to the following January?

Unit 2   Symbols

Try the following test:

1   In each of the following algebraic expressions identify the variables
and distinguish the terms and their coefficients:
(a) $p - 5q + 3r$ (b) $st - 2s/t + 3t$
(c) $uv/4w - 3u/w + 2v/w + uvw$ (d) $7a/bc - 6b/ac + 5c/ab$

Unknowns

In the previous unit we saw how to use a letter of the alphabet to represent an
unknown number. In the problem that we considered the letter represented just one
unknown number but in many other problems we require a symbolism that can be
used to represent any one of a collection of numbers. Central to this symbolism is the
notion of a **variable**.

Placeholders

Arithmetic deals with numbers, or to be more correct, it deals with numerals and their
manipulation when combined with the various arithmetic operations. Throughout
our discussion of the development of the real number system the notion of a
mathematical system was continually being stressed. To re-cap, a mathematical
system consists of:

■ *a collection of objects*
■ *a collection of operations between objects*
■ *a collection of rules obeyed when objects and operations are combined.*

In arithmetic the objects are the real numbers, represented by the numerals and the
operations are addition, subtraction, multiplication, division and raising to a power.
One of the rules obeyed by two real numbers when combined under addition is that
the order of addition does not matter. For example:

$1 + 3 = 3 + 1$
$2 + 3 = 3 + 2$
$3 + 3 = 3 + 3$
. . . .

This sequence of equalities exemplifies the rule. It is telling us that it makes no
difference if we add 3 to a given number or we add the given number to 3, the result
is just the same. We can generalize this by writing:

■ $+ 3 = 3 + $ ■

where the symbol ■ is called a **placeholder** in the sense that it holds the place where we can insert any real number and maintain the truth of the equality. This notation, however, does pose a problem. For example, if we were to extend our discussion to the rule concerning the use of brackets to associate two or more real numbers:

$$(■ + ■) + 3 = ■ + (■ + 3)$$

the immediate question posed is:

Must it be the same number in all four places?

To avoid questions like this arising we need a collection of different symbols to use as placeholders when different numbers are to be inserted. The collection that we use consists of letters of the alphabet. For example, we can change the symbol ■ to x and write the first rule as:

$x + 3 = 3 + x$ where x is any given real number.

For the second rule, we require two different symbols. We use x and y:

$(x + y) + 3 = x + (y + 3)$ where x and y are both real numbers.

Both of these rules can be extended to give their most general form as:

$x + y = y + x$ where both x and y are real numbers.
$(x + y) + z = x + (y + z)$ where x, y and z are any three real numbers.

Indeed, all the rules of arithmetic that we have already met by specific example can be generalized in this way. We shall say more of this later.

Operations
All the operations of arithmetic that act between numbers when they are represented by their numerals also act between numbers when they are represented by variables. Thus we speak of the:

sum of x and y	written as	$x + y$
difference between x and y	written as	$x - y$
product of x and y	written as	xy
quotient of x and y	written as	x/y or $x + y$
raising x **to the power** y	written as	x^y

Notice that the product of x and y is written as:

xy and not as $x \times y$.

This avoids any confusion between the symbols $\times$ and x.

Algebraic expressions

An algebraic expression consists of a collection of numerals and variables linked together with the various arithmetic operations. For example:

$$5s - 2st$$

is an algebraic expression in the two variables s and t. Each component of the expression is called a **term** of the expression. Here we see that there are two terms, namely:

$$5s \text{ and } -2st$$

The numerals involved in each term are called the **coefficients** of their respective terms. So that:

5 is the coefficient of the s term and -2 is the coefficient of the st term.

Worked Examples

5.4 In each of the following algebraic expressions identify the variables and distinguish the terms and their coefficients:
  (a) $x + 2y - 3z$ (b) $ab - a/c + bc - 3ab/c$
  (c) $prq + 4prs - rq - 5pr$ (d) $4xyz - 2x/y + 5/x$

Solution:
(a) $x + 2y - 3z$
  There are three terms in the variables x, y and z in this algebraic expression:

  x, $2y$ and $-3z$

  Their respective coefficients are:

  1, 2 and -3.

  Notice that the coefficient of the x term is 1.

(b) $ab - a/c + bc - 3ab/c$

  There are four terms in the variables a, b and c in this algebraic expression:

  ab, $-a/c$, bc and $-3ab/c$

  with respective coefficients:

  1, -1, 1 and -3.

(c) $prq + 4prs - rq - 5pr$

There are four terms in the variables p, q, r and s in this algebraic expression:

prq, $4prs$, $-rq$ and $-5pr$

with respective coefficients:

1, 4, -1 and -5

(d) $4xyz - 2x/y + 5/x$

There are three terms in the variables x, y and z in this algebraic expression:

$4xyz$, $-2x/y$ and $5/x$

with respective coefficients:

4, -2 and 5

Exercises

5.4 In each of the following algebraic expressions identify the variables and distinguish the terms and their coefficients:
(a) $s + 5t - 7u$ (b) $x/y + 5xy - 2x$
(c) $9abc/d + 12ab/c + 11c/ab - ad$ (d) $5uv/w - 5u/w + 7/w + 4uv$

Unit 3   Into algebra

Try the following test:

1   Eliminate the brackets in each of the following algebraic expressions:
    (a) $4(a + 2)$ (b) $p(q - 3)$
    (c) $x(y - 5z)$ (d) $2(x + y) - 5(x + y)$
    (e) $g(h - k) - g(h + k)$ (f) $2u(v - w) - 7w(v + u)$
    (g) $2p(3q + 4(r - s))$

2   Reduce each of the following expressions to their simplest form:
    (a) $x^2y^3x^3y^{-5}$ (b) $(p^3q^2)/(q^{-3}p^4)$
    (c) $(a^3b^2c)/(ab^2c^3)$ (d) $(u^{-1/2}v^{2/3})^6$
    (e) $(g^{-3})^{1/6}(h^6)^{-1/3}(g/h^2)^{-1/4}$ (f) $((x^{2/3})^{1/2}(y^{4/5})^{-1/4})^{15/2}$

The rules of algebra
We have already seen how the arithmetic operations and the numerals combine according to the rules of arithmetic. When we looked at a particular rule we stated that it was generally true but we exemplified it by a specific instance. Here, we list these rules for the first four operations in their general form using algebraic symbolism.

Order of adding or multiplying
The order in which we add or multiply two numbers does not matter:

$x + y = y + x$ for all real numbers x and y
$xy = yx$ for all real numbers x and y

Notice that this is not true for subtraction and division:

$x - y \neq y - x$ for all real numbers x and y
$x/y \neq y/x$ for all real numbers x and y

Whilst it may be true, for example, that:

$x - y = y - x$ when $x = y$,

it is a special case and not true for *all* pairs of real numbers.

Brackets
Brackets do not affect the addition or multiplication of groups of numbers:

$x + (y + z) = (x + y) + z = x + y + z$

$$x(yz) = (xy)z = xyz$$

Multiplying through

When a number multiplies a sum or difference of two or more numbers in brackets the brackets can be removed by multiplying through.

From the left:

$$x(y + z) = xy + xz$$
$$x(y - z) = xy - xz$$

From the right:

$$(x + y)z = xz + yz$$
$$(x - y)z = xz - yz$$

Division requires special care:

$$(x + y) \div z = \frac{x + y}{z} = \frac{x}{z} + \frac{y}{z}$$

whereas:

$$x \div (y + z) = \frac{x}{y + z} \neq \frac{x}{y} + \frac{x}{z}$$

Take care with this and the following rule because they both represent commonly made mistakes.

Also

$$(x - y) \div z = \frac{x}{z} - \frac{y}{z}$$

whereas:

$$x \div (y - z) = \frac{x}{y - z} \neq \frac{x}{y} - \frac{x}{z}$$

Identity

The number 0 is called the **identity** element for addition and subtraction as it leaves the number to which it adds or from which it subtracts unaltered:

$$x + 0 = x \text{ and } x - 0 = x$$

Similarly the number 1 is called the identity element for multiplication and division:

$$x.1 = x \text{ and } x \div 1 = x$$

Note: Division by 0 is not allowed because it cannot be defined.

Worked Examples

5.5 Eliminate the brackets in each of the following algebraic expressions:
  (a) $5(x + 3)$ (b) $x(2 - y)$
  (c) $s(t + u)$ (d) $3(x - y) + 2(x + y)$
  (e) $(x + y) - (x - y)$ (f) $x(y - z) - x(z + y)$
  (g) $p(3 - 4(q + r(2 + s)))$

Solution:
(a) $5(x + 3) = 5x + 15$

(b) $x(2 - y) = 2x - xy$

  Notice that $x2 = 2x$ (we always put the coefficient before the variable).

(c) $s(t + u) = st + su$

(d) $3(x - y) + 2(x + y) = 3x - 3y + 2x + 2y$

(e) $(x + y) - (x - y) = x + y - x + y$

  Notice that:

  $-(x - y) = -1(x - y) = (-1)x - (-1)y = -x + y$

(f) $x(y - z) - x(z + y) = xy - xz - xz - xy$

(g) When eliminating brackets within brackets (nested brackets) we eliminate the innermost bracket first.

  $$p(3 - 4(q + r(2 + s))) = p(3 - 4(q + 2r + rs))$$
  $$= p(3 - 4q - 8r - 4rs)$$
  $$= 3p - 4pq - 8pr - 4prs$$

  Notice that in each term throughout these examples we have written the cofficient first, followed by the variables in alphabetic order. Whilst this is not strictly essential it does provide a consistent approach to the written algebra.

Exercises

5.5 Eliminate the brackets in each of the following algebraic expressions:
  (a) $3(x + 4)$ (b) $s(6 - t)$
  (c) $p(q + 3r)$ (d) $4(u - v) - 8(u + v)$
  (e) $(f - g) + (f + g)$ (f) $k(m - n) - m(k + n)$
  (g) $a(2 + 3(b - c(9 - d)))$

Raising to a power

The fifth arithmetic operation is raising to a power and the rules that govern this operation can be generalized as follows:

Given any real number a:

Power unity
$a^1 = a$

Power zero
$a^0 = 1$

Addition of powers
$a^x a^y = a^{x+y}$ where x and y are any pair of real numbers

Notice also that:

$a^x b^x = (ab)^x$

Negative powers
$a^{-x} = 1/a^x$ where x is any real number

Subtraction of powers
$a^x/a^y = a^{x-y}$ where x and y are any pair of real numbers

Multiplication of powers
$(a^x)^y = a^{xy}$ where x and y are any pair of real numbers

Fractional powers
$a^{1/n}$ is the nth root of a where n is any natural number

Worked Examples

5.6 Reduce each of the following expressions to their simplest form:

(a) $p^3 q^4 p^{-2} q^2$　　　　　　　　　(b) $(a^2 b^3)/(ab^4)$
(c) $(x^2 y z^{-3})/(x^{-1} y^2 z)$　　　　　(d) $(s^{3/2} t^{1/4})^6$
(e) $(u^2)^{1/3}(v^{-4})^{1/3}(u/v)^{-1/3}$　　(f) $((a^{1/3})^2 (b^{-1/3})^{-2})^{3/2}$

Solution:

(a) $p^3 q^4 p^{-2} q^2 = p^3 p^{-2} q^4 q^2$
$\qquad\qquad\quad = p^{3-2} q^{4+2}$
$\qquad\qquad\quad = pq^6$

(b) $(a^2 b^3)/(ab^4) = (a^2 b^3)(ab^4)^{-1}$
$\qquad\qquad\quad = (a^2 b^3)(a^{-1} b^{-4})$
$\qquad\qquad\quad = a^2 b^3 a^{-1} b^{-4}$

$$= a^2 a^{-1} b^3 b^{-4}$$
$$= a^{2-1} b^{3-4}$$
$$= ab^{-1}$$
$$= a/b$$

(c) $(x^2 yz^{-3})/(x^{-1}y^2 z) = (x^2 yz^{-3})(x^{-1}y^2 z)^{-1}$
$$= x^2 yz^{-3} x^1 y^{-2} z^{-1}$$
$$= x^2 x^1 yy^{-2} z^{-3} z^{-1}$$
$$= x^3 y^{-1} z^{-4}$$

(d) $(s^{3/2} t^{1/4})^6 = (s^{3/2})^6 (t^{1/4})^6$
$$= (s^{18/2})(t^{6/4})$$
$$= s^9 t^{3/2}$$

(e) $(u^2)^{1/3}(v^{-4})^{1/3}(u/v)^{-1/3} = u^{2/3} v^{-4/3} u^{-1/3} v^{1/3}$
$$= u^{2/3} u^{-1/3} v^{-4/3} v^{1/3}$$
$$= u^{1/3} v^{-3/3}$$
$$= (u^{1/3})/v$$

(f) $((a^{1/3})^2 (b^{-1/3})^{-2})^{3/2} = ((a^{2/3})(b^{2/3}))^{3/2}$
$$= ((ab)^{2/3})^{3/2}$$
$$= ab$$

Exercises

5.6 Reduce each of the following expressions to their simplest form:

(a) $a^{-1} b^5 a^2 b^{-4}$

(b) $(u^4 v)/(v^{-3} u^{-2})$

(c) $(p^{-2} q^3 r^2)/(p^{-2} q^{-1})$

(d) $(x^{-1/3} y^{5/6})^{12}$

(e) $(s^{-2})^{1/2}(t^4)^{1/2}(t/s)^{-3/4}$

(f) $((x^{-3/4})^{1/3}(y^{3/8})^{-1/3})^{8/5}$

Module 5   Further exercises

1   If I buy six cans of dog food and four cans of cat food I will have to pay £6.70. However, if I buy four cans of dog food and six cans of cat food I will only have to pay £6.30. How much does each can of dog food and cat food cost?

2   Beautiful maiden with beaming eyes, tell me, as thou understandst the right method of inversion, which is the number multiplied by 3, then increased by 3/4 of the product, then divided by 7, diminished by 1/3 of the quotient, multiplied by itself, diminished by 52, by the extraction of the square root, addition of 8, and division by 10 gives the number 2?

3   The eighteenth century astronomer Bode found a pattern in the distances of the planets from the sun. Taking the distance between Mercury and the sun as 4 units Bode listed his numbers as follows:

Planet	Bode's numbers		Actual (to nearest whole number)
Mercury	$0 + 4 =$	4	4
Venus	$3 + 4 =$	7	7
Earth	$6 + 4 =$	10	10
Mars	$12 + 4 =$	16	15
Jupiter	$48 + 4 =$	52	52
Saturn	$* + 4 =$	100	95
Uranus	$* + 4 =$	*	192
Neptune	$* + 4 =$	*	301
Pluto	$* + 4 =$	*	*

Complete the table. Bode's numbers only went as far as Saturn's distance because the outer three planets were not discovered until after his death.

4   In each of the following algebraic expressions identify the variables and distinguish the terms and their coefficients:
(a) $2a - 3b + 4c$ (b) $3pq - 2p/q + 4q/p$
(c) $7xy/z - 5xz/y$ (d) $2uv/w + 4u/w - 1/w + 2uv$

5   Eliminate the brackets in each of the following algebraic expressions:
(a) $7(2 - u)$ (b) $a(b - 3)$
(c) $u(v - 2w)$ (d) $3(a - 2b) - 2(b - 3a)$
(e) $-(a - b) - (b - a)$ (f) $2x(3y - 4z) - 3y(2x - 4z)$
(g) $2x(3y - 4(z - 2))$

130

6   Reduce each of the following expressions to their simplest form:

(a) $u^5 v^{-3} v^{-1} u^2$

(b) $(x^2 y^{-2})/(x^{-3} y^{-3})$

(c) $(r^{-3} s^{-1} t^2)/(s^{-1} r^{-4})$

(d) $(a^{-1/3} b^{3/2})^{-8}$

(e) $(p^{-1})^{-1/3}(q^2)^{1/4}(q/p^2)^{-3/5}$

(f) $((m^{-7/8})^{1/7}(n^{3/14})^{-1/3})^{7/11}$

Module 6

Algebraic manipulation

OBJECTIVES

When you have completed this chapter you will be able to:

- Identify and combine like terms in an algebraic expression

- Identify and factorize similar terms in an algebraic expression

- Expand factored expressions

- Multiply two algebraic expressions together

- Use standard factorizations

- Manipulate and simplify algebraic fractions

There are five Units in this module:

Unit 1: Early algebraic skills
Unit 2: Algebraic expressions
Unit 3: Multiplying and factorizing
Unit 4: Pascal's triangle
Unit 5: Fractions

Unit 1   Early algebraic skills

Try the following test:

1   Combine like terms in each of the following algebraic expressions:
   (a) $-11zx + 5yz + 2xz - 3xy - 4zy + 2y$
   (b) $4s + 7st - 2t - 8ts$
   (c) $9lmn - 3mn - 5mnl + 8nlm - 3nm$
   (d) $5x^2y - 6x + 3yx^2 - 5y^2x - 6x + 3y^2x$
   (e) $9s^2t^4 - 3t^4s^2 - t^3s + 2ts^3 + 7st + 4ts - 6st^3 + 4s^3t$

2   Factorize similar terms in each of the following algebraic expressions:
   (a) $-12mn - 6lm$ (b) $-18xy + 4xz + 9wy - 2wz$
   (c) $5abc - 10bc + 15ab - 10b$ (d) $4x^2y - 8y^2x + 6xy - 12x^2y^2$

Basic manipulative skills

Algebraic manipulation involves writing algebraic expressions in alternative forms. This may mean writing an algebraic expression in an alternative form that has fewer terms or, conversely, one that has more terms than the original expression. In either case, to do this effectively requires a facility in the basic skills of manipulating symbols. In this chapter we shall be concerned with the basic skills of simplifying and, alternatively, the expanding of algebraic expressions. To become proficient in this skill requires practice and to assist you in gaining the necessary practice a number of exercises follow each displayed skill.

Combining or collecting like terms

The first skill to achieve when handling algebraic expressions is the ability to recognize like terms and to differentiate between unlike terms. For example, in the expression:

$$ab - cb + 2ac - 4ab + 3bc$$

the terms:

ab and $4ab$ are considered to be **like terms** because they both contain the same product of symbols ab

and:

ab and cb are considered to be **unlike** terms because they do not both contain the same product of symbols. They do, however, both contain the symbol b but we shall discuss this aspect later.

Notice:

Because $cb = bc$ we can say that cb and $3bc$ are like terms.

The first step when dealing with an algebraic expression is to rewrite it so that all like terms are adjacent to each other. We rewrite the above expression as:

$ab - 4ab - bc + 3bc + 2ac$

Notice that we have replaced cb by bc so that the second pair of like terms look similar.

The next step is to combine like terms by executing the appropriate operations. This means that the expression now becomes:

$-3ab + 2bc + 2ac$

An alternative name for this process is **collecting** like terms together.

Worked Examples

6.1 Combine like terms in each of the following algebraic expressions:
   (a) $3ab + 5ac - 8bc - 2ac + ba$
   (b) $-5s + 4st + 7s - 9ts$
   (c) $8xyz + 5yxz + 2zy - 3xz + 4yz$
   (d) $15p^2q - 3p - 7qp^2 + 12p^2r - 6r + 2r^2p$
   (e) $-3uv^3w - 5v^3u + 2wuv^3 - 7v^3u + 8uv^3w$
        $- 5uw - uv^3 + 3wuv^3 - 2uw - 3uv^3$

Solution:
(a) $3ab + 5ac - 8bc - 2ac + ba = 3ab + ba + 5ac - 2ac - 8bc$
$$= 3ab + ab + 3ac - 8bc$$
$$= 4ab + 3ac - 8bc$$

(b) $-5s + 4st + 7s - 9ts = -5s + 7s + 4st - 9st$
$$= 2s - 5st$$

(c) $8xyz - 5yxz + 2zy - 3xz + 4yz = 8xyz + 5xyz + 2yz + 4yz - 3xz$
$$= 13xyz + 6yz - 3xz$$

(d) $15p^2q - 3p - 7qp^2 + 12p^2r - 6r + 2r^2p$
$$= 15p^2q - 7p^2q - 3p + 12p^2r - 6r + 2r^2p$$
$$= 8p^2q - 3p + 12p^2r - 6r + 2r^2p$$

(e) $-3uv^3w - 5v^3u + 2wuv^3 - 7v^3u + 8uv^3w - 5uw - uv^3 + 3wuv^3 - 2uw - 3uv^3$
$$=-3uv^3w + 2uv^3w + 8uv^3w - 5uv^3 - 7uv^3 - 3uv^3 - uv^3 - 3uv^3 - 5uw - 2uw$$
$$= 10uv^3w - 16uv^3 - 7uw$$

Exercises

6.1 Combine like terms in each of the following algebraic expressions:
- (a) $7jk + 4km - 3jm - kj + 2mj$
- (b) $6d - 2de - e + 3ed$
- (c) $-4pqr + 7qpr + rpq + 2qr - 7pr$
- (d) $11c^3d + 4c - 5dc^3 + 9c - 6d + 7d^3c$
- (e) $6lm^3 + 5m^3n - 3l^3n - 5lmn^3 + 3nm^3 - lm^3 + 8l^3n - 2n^3ml$

Brackets

In algebra, just as in the arithmetic of the numerals, we use the BEODMAS precedence rules to dictate the order in which operations are executed. For example, in the expression:

$$ac - 3bc + 2b \text{ or } ac - 3bc/2b$$

the precedence rules dictate that the division is performed before the subtraction. To force the subtraction to be performed before the division we must, as in arithmetic, resort to using brackets. Thus, in the expression:

$$(ac - 3bc) + 2b \text{ or } (ac - 3bc)/2b$$

the subtraction is executed before the division. In actuality, the precedence rules still hold sway because they tell us that brackets are evaluated before division and executing the bracket means performing the subtraction that is within it.

Factors and similar terms

The bracketed portion of the expression that we have just considered is:

$$(ac - 3bc)$$

Here the symbol c is common to both terms – that is, it is a **common factor**. Accordingly, the factor c can be **factored out** and the bracketed expression written as:

$$c(a - 3b)$$

Terms which contain factors in common are called **similar terms** and extracting common factors in this way is a process referred to as **factorizing siimilar terms**.

Worked Examples

6.2 Factorise similar terms in each of the following algebraic expressions:

- (a) $5pq + 3pr$
- (b) $8uv - 4vw + 2uw - 4w^2$
- (c) $-3abc + 9bc + 6ac - 18b$
- (d) $14j^2kl + 7j^2l - 6kj - 4j$

Solution:

(a) $5pq + 3pr = p(5q + 3r)$

(b) $8uv - 4vw + 2uw - 4w^2 = 4v(u - w) + 2w(u - w)$
$$= (u - w)(4v - 2w)$$
$$= 2(u - w)(2v - w)$$

Note that you can factor in different ways:

$$8uv - 4vw + 2uw - 4w^2 = 8uv - 2w(2v - u + 2w)$$

(c) $-3abc + 9bc + 6ab - 18b = 3bc(-a + 3) + 6b(a - 3)$
$$= -3bc(a - 3) + 6b(a - 3)$$
$$= (a - 3)(6b - 3bc)$$
$$= (a - 3)3b(2 - c)$$
$$= 3b(a - 3)(2 - c)$$

(d) $14j^2kl + 7j^2l - 6kj - 4j = 7j^2l(2k + 1) - 2j(3k + 2)$
$$= j(7jl(2k + 1) - 2(3k + 2))$$

Exercises

6.2 Factorize similar terms in each of the following algebraic expressions:

(a) $4ab - 2bc$　　　　　　　　　　(b) $9pq - 3pr + 3sq - sr$

(c) $-2uvw + 4vw + 3uv - 6v$　　　(d) $8l^2mn + 2ln - 12im - 3$

Unit 2   Algebraic expressions

Try the following test:

1   Simplify each of the following algebraic expressions:
   (a) $-xz - 2xy + 6zx + 4yx$
   (b) $-4r - 8sr - t - 2st$
   (c) $9jkl + 6ljk + 7klj - 5jk - lk - 3kl$
   (d) $4u^2v - 4v - 2vu^3 + 8w - 4u^2w$
   (e) $-5rs^3 + 5s^3r - 2t^3r + 6t^2r^2 + rs^3 - ts^3 - t^3r - 3r^2t^2$

2   Expand each of the following algebraic expressions:
   (a) $-x(3y + 4x)$ (b) $-9p(-3q + 4r)$
   (c) $(-6a + 3b)3c$ (d) $-7u(5v + 2u) - 8v(-3v - 5u)$
   (e) $-(-8a + 3b)(-4c) + 7b(-3b - 7a)$

Simplifying algebraic expressions

By combining like terms and factorizing common factors in similar terms we can often reduce the number of terms in an expression. This is referred to as **simplifying the expression**. For example, the expression:

$$ab - cb + 2ac + 4ab + 3bc$$

can be simplified by combining like terms to produce:

$$2bc + 5ab + 2ac$$

The first two terms of this expression can now be factorized to yield:

$$b(2c + 5a) + 2ac$$

or, alternatively, by factorizing the second two terms to yield:

$$2bc + a(5b + 2c)$$

or even:

$$c(2b + 2a) + 5ab$$

The last factorization can be further factorized to yield:

$$c(2(b + a)) + 5ab = 2c(b + a) + 5ab$$

Which of the three different factorizations to select will depend upon the problem and what procedures are to follow the factorization process. Here, factorizing was all that was required so any one of the three possible options is correct.

Worked Examples

6.3 Simplify each of the following algebraic expressions:
   (a) $7jk - 3jm - kj + 2mj$
   (b) $6d - 2de - e + 3ed$
   (c) $-4pqr + 7qpr + rpq + 2qr - 7pr$
   (d) $11cd + 4c - 7dc + 9b + 9db$
   (e) $6lm^3 - 5m^3n - 3ln - 5n^2 - 2nm^3 - lm^3 + 8ln - 2n^2$

Solution:
(a) $7jk - 3jm - kj + 2mj = 6jk - mj$
$$= j(6k - m)$$

(b) $6d - 2de - e + 3ed = 6d - e + ed$
$$= 6d - e(1 - (d) \ or = d(6 + (e) - e$$

(c) $-4pqr + 7qpr + rpq + 2qr - 7pr = 4pqr + 2qr - 7pr$
$$= r(4pq + 2q - 7p)$$
$$= r(p(4q - 7) + 2q)$$

(d) $11cd + 4c - 7dc + 9b + 9db = 4cd + 4c + 9b + 9db$
$$= 4c(d + 1) + 9b(1 + (d)$$
$$= 4c(d + 1) + 9b(d + 1)$$
$$= (4c + 9b)(d + 1)$$

(e) $6lm^3 - 5m^3n - 3ln - 5n^2 - 2nm^3 - lm^3 + 8ln - 2n^2 = 5lm^3 - 7m^3n + 5ln - 7n^2$
$$= m^3(5l - 7n) + n(5i - 7n)$$
$$= (m^3 + n)(5i - 7n)$$

Exercises

6.3 Simplify each of the following algebraic expressions:
   (a) $13pq + 8pr - 5qp - 3rp$ (b) $9a + 3ab - 3c - bc$
   (c) $-15xyz - 4zxy + 3yzx + 8xy - zy$ (d) $15p^2r^2 - 10p^2s - 24qr^3 + 16qs$
   (e) $2ca^2 - 2a^2b - 2abc + 2ab^2$

Multiplying factors

Just as we extracted common factors so we can reverse the process and multiply them out. For example, the factored expression:

$q(p - 7rs)$

can be multiplied out to convert it to the expression:

$qp - 7qrs$

This process of multiplying out common factors is known as *expanding* the expression and is the reverse process to simplifying an expression.

Worked Examples

6.4 Expand each of the following algebraic expressions:

(a) $a(3b + 4c)$ (b) $5p(q - 3r)$

(c) $(2u - 3v)w$ (d) $3a(5a + 7b) + 4b(6a - 9b)$

(e) $(x - 2y)3z - 5y(2z - 3x)$

Solution:

(a) $a(3b + 4c) = 3ab + 4ac$

(b) $5p(q - 3r) = 5pq - 15pr$

(c) $(2u - 3v)w = 2uw - 3vw$

(d) $3a(5a + 7b) + 4b(6a - 9b) = 15a^2 + 21ab + 24ab - 36b^2$
$$= 15a^2 + 45ab - 36b^2$$

(e) $(x - 2y)3z - 5y(2z - 3x) = 3xz - 6yz - 10yz + 15yx$
$$= 3xz - 16yz + 15yx$$

Exercises

6.4 Expand each of the following algebraic expressions:

(a) $r(2s + 5t)$ (b) $7a(2b - 9c)$

(c) $(4p - 5q)r$ (d) $12x(3x - 4y) - y(8x - 7y)$

(e) $(5j - 6k)2m - 3k(4m + 2j)$

Unit 3   Multiplying and factorizing

Try the following test:

1   Multiply out each of the following algebraic expressions:
   (a) $(7u - 3v)(5w - 9x)$ (b) $(-7r - 3s)(2r + 3s)$
   (c) $(3x^2 + 2)(5x^2 + 3x - 1)$ (d) $(5x - 3)(2 - 5x - 6x^2 - 3x^3)$
   (e) $(4 - 3x - x^2)(5x^3 - 2x + 4)$

2   Factorize each of the following algebraic expressions:
   (a) $x^2 - 7x + 10$ (b) $x^2 + 6x + 9$
   (c) $x^2 - 64$ (d) $x^3 - 1331$
   (e) $x^3 + 216$ (f) $4x^2 - 4x - 80$
   (g) $x^2 + 3x - 10$ (h) $9x^2 - 4$
   (j) $1331x^3 - 343$ (k) $729x^3 + 125$

Multiplication of expressions

The process of multiplying out factors can be straightforwardly extended to multiplying any two algebraic expressions. For example:

$$
\begin{aligned}
(2x - 5y)(3a + 6b) &= 2x(3a - 6b) - 5y(3a + 6b) \\
&= 6xa - 12xb - 15ya - 30yb \\
&= 6ax - 12bx - 15by - 30by
\end{aligned}
$$

Notice that in the last line of this equation we have rewritten the previous line with the symbols in alphabetical order. Whilst it is not strictly necessary to do this it does assist in the identification of like and similar terms.

When multiplying expressions involving powers of a variable all similar powers are accumulated together and the final result is written in ascending or descending order of powers of the variable. For example:

$$
\begin{aligned}
(x^2 - 3)(x^3 + 2x^2 - 5x + 9) &= x^2(x^3 + 2x^2 - 5x + 9) - 3(x^3 + 2x^2 - 5x + 9) \\
&= x^5 + 2x^4 - 5x^3 + 9x^2 - 3x^3 - 6x^2 + 15x - 27 \\
&= x^5 + 2x^4 - 8x^3 + 3x^2 + 15x - 27
\end{aligned}
$$

Worked Examples

6.5 Multiply out each of the following algebraic expressions:
   (a) $(2a + 3b)(c - 5d)$ (b) $(3x - 2y)(2x + 3y)$
   (c) $(x^2 + 2)(x^2 + 3x - 4)$ (d) $(3x - 5)(x^3 - x^2 + x - 1)$
   (e) $(5x^2 - 4x + 3)(2x^3 - x^2 + x - 4)$

Solution:

(a) $(2a + 3b)(c - 5d) = 2a(c - 5d) + 3b(c - 5d)$
$$= 2ac - 10ad + 3bc - 15bd$$

(b) $(3x - 2y)(2x + 3y) = 3x(2x + 3y) - 2y(2x + 3y)$
$$= 6x^2 + 9xy - 4yx - 6y^2$$
$$= 6x^2 + 5xy - 6y^2$$

(c) $(x^2 + 2)(x^2 + 3x - 4) = x^2(x^2 + 3x - 4) + 2(x^2 + 3x - 4)$
$$= x^4 + 3x^3 - 4x^2 + 2x^2 + 6x - 8$$
$$= x^4 + 3x^3 - 2x^2 + 6x - 8$$

(d) $(3x - 5)(x^3 - x^2 + x - 1) = 3x(x^3 - x^2 + x - 1) - 5(x^3 - x^2 + x - 1)$
$$= 3x^4 - 3x^3 + 3x^2 - 3x - 5x^3 + 5x^2 - 5x + 5$$
$$= 3x^4 - 8x^3 + 8x^2 - 8x + 5$$

(e) $(5x^2 - 4x + 3)(2x^3 - x^2 + x - 4)$
$$= 5x^2(2x^3 - x^2 + x - 4) - 4x(2x^3 - x^2 + x - 4) + 3(2x^3 - x^2 + x - 4)$$
$$= (10x^5 - 5x^4 + 5x^3 - 20x^2) - (8x^4 - 4x^3 + 4x^2 - 16x) + (6x^3 - 3x^2 + 3x - 12)$$
$$= 10x^5 - 5x^4 + 5x^3 - 20x^2 - 8x^4 + 4x^3 - 4x^2 + 16x + 6x^3 - 3x^2 + 3x - 12$$
$$= 10x^5 - 13x^4 + 15x^3 - 27x^2 + 19x - 12$$

Exercises

6.5 Multiply out each of the following algebraic expressions:

(a) $(4p + 2q)(3r - s))$ (b) $(9a - 3b)(2a + 4b)$
(c) $(x^2 - 3)(4x^2 - x + 2)$ (d) $(x + 7)(2x^3 + 5x^2 - 3x + 7)$
(e) $(2x^2 + 3x - 1)(3x^3 + x^2 - 2x - 6)$

Standard factorizations

The product of certain special types of expressions gives rise to an easily identifiable factorization. For example, the product of the two expressions $(x + 2)$ and $(x + 3)$ yields:

$$(x + 2)(x + 3) = x(x + 3) + 2(x + 3)$$
$$= x^2 + (2 + 3)x + (2 \times 3)$$
$$= x^2 + 5x + 6$$

Notice that the coefficent of the x-term is the **sum** $2 + 3$ and that the third term is the **product** 2×3. Consequently, to factorize the expression $x^2 + 3x + 2$ we write:

$$x^2 + 3x + 2 = (x + a)(x + b)$$

and look for two number a and b where $a + b = 3$ and $ab = 2$. By trying different pairs of numbers, that is by trial and error, we find that the two numbers are:

$a = 2$ and $b = 1$

giving the factorization as:

$x^2 + 3x + 2 = (x + 2)(x + 1)$

The general form of this expression and its factorization is:

$(x + a)(x + b) = x^2 + (a + b)x + ab$

which is referred to as a **standard factorization**. Other standard factorizations that are commonly met are:

$(x + a)(x + a) = x^2 + 2ax + a^2$ **repeated factors** where $b = a$.

$(x + a)(x - a) = x^2 - a^2$ **the difference of two squares** where $b = -a$.

$(x + a)(x^2 - ax + a^2) = x^3 + a^3$ **the sum of two cubes.**

$(x - a)(x^2 + ax + a^2) = x^3 - a^3$ **the difference of two cubes.**

The latter two factorizations can easily be demonstrated to be correct by multiplying out the factors.

Worked Examples

6.6 Factorize each of the following algebraic expressions:
   (a) $x^2 + 4x + 3$ (b) $x^2 - 5x + 6$
   (c) $x^2 - 4$ (d) $x^3 - 27$
   (e) $x^3 + 125$ (f) $3x^2 + 21x + 36$
   (g) $x^2 + x - 2$ (h) $4x^2 - 9$
   (i) $8x^3 - 64$ (j) $27x^3 + 8$

Solution:
(a) $x^2 + 4x + 3 = (x + a)(x + b)$

Here $a + b = 4$ and $ab = 3$ so $a = 1$ and $b = 3$ giving:

$x^2 + 4x + 3 = (x + 1)(x + 3)$

(b) $x^2 - 5x + 6 = (x + a)(x + b)$

Here $a + b = -5$ and $ab = 6$ so $a = -2$ and $b = -3$ giving:

$x^2 - 5x + 6 = (x - 2)(x - 3)$

(c) $x^2 - 4 = x^2 - 2^2$ ⠀⠀⠀⠀⠀A difference of two squares. Therefore:

$x^2 - 4 = (x + 2)(x - 2)$

(d) $x^3 - 27 = x^3 - 3^3$ ⠀⠀⠀⠀A difference of two cubes. Therefore:

$$x^3 - 27 = (x - 3)(x^2 + 3x + 3^2)$$
$$= (x - 3)(x^2 + 3x + 9)$$

(e) $x^3 + 125 = x^3 + 5^3$ ⠀⠀⠀⠀A sum of two cubes. Therefore:

$x^3 + 125 = (x + 5)(x^2 - 5x + 25)$

(f) $3x^2 + 21x + 36$ ⠀⠀⠀⠀⠀Here each term has a common factor 3 so we factorize that out first to yield:

$$3x^2 + 21x + 36 = 3(x^2 + 7x + 12)$$
$$= 3(x + 3)(x + 4)$$

(g) $x^2 + x - 2 = (x + a)(x + b)$

Here $a + b = 1$ and $ab = -2$ so $a = -1$ and $b = 2$ giving:

$x^2 + x - 2 = (x - 1)(x + 2)$

(h) $4x^2 - 9 = (2x)^2 - 3^2$ ⠀⠀A difference of two squares. Therefore:
$$= (2x + 3)(2x - 3)$$

(i) $8x^3 - 64 = (2x)^3 - 4^3$ ⠀⠀A difference of two cubes. Therefore:
$$= (2x - 4)((2x)^2 + (2x)4 + 4^2)$$
$$= (2x - 4)(4x^2 + 8x + 16)$$

(j) $27x^3 + 8 = (3x)^3 + 2^3$ ⠀⠀A sum of two cubes. Therefore:
$$= (3x + 2)((3x)^2 - (3x)2 + 2^2)$$
$$= (3x + 2)(9x^2 - 6x + 8)$$

Exercises

6.6 Factorize each of the following algebraic expressions:

(a) $x^2 + 7x + 10$ ⠀⠀⠀⠀⠀⠀⠀⠀(b) $x^2 - 4x + 4$
(c) $x^2 - 25$ ⠀⠀⠀⠀⠀⠀⠀⠀⠀⠀⠀(d) $x^3 - 1$
(e) $x^3 + 1000$ ⠀⠀⠀⠀⠀⠀⠀⠀⠀(f) $5x^2 + 5x - 30$
(g) $x^2 - x - 2$ ⠀⠀⠀⠀⠀⠀⠀⠀⠀(h) $16x^2 - 9$
(j) $27x^3 - 1$ ⠀⠀⠀⠀⠀⠀⠀⠀⠀⠀(k) $125x^3 + 64$

Unit 4   Pascal's triangle

Binomial expansions

A **binomial** is a sum of two numbers and we have already seen that the square of the binomial $(a + b)$ is given as:

$$(a + b)^2 = (a + b)(a + b)$$
$$= a(a + b) + b(a + b)$$
$$= a^2 + 2ab + b^2$$

The expression on the right-hand side of this equation is known as the **expansion** of the expression on the left-hand side. We can extend this to find the expansion of the cube of the binomial $(a + b)^3$:

$$(a + b)^3 = (a + b)(a + b)^2$$
$$= (a + b)(a^2 + 2ab + b^2)$$
$$= a(a^2 + 2ab + b^2) + b(a^2 + 2ab + b^2)$$
$$= a^3 + 2a^2b + ab^2 + ba^2 + 2ab^2 + b^3$$
$$= a^3 + 3a^2b + 3ab^2 + b^3$$

Notice that in this expansion the powers in each term add up to 3 and the terms are listed in the strict order of decreasing power of a.

Likewise, the expansion of the fourth power of the binomial can be shown to be:

$$(a + b)^4 = a^4 + 4a^3b + 6a^2b^2 + 4ab^3 + b^4$$

Here, the powers in each term now add up to 4 and the terms are listed in the strict order of decreasing power of a.

In each of these three expansions you will note a certain symmetry. Listed in order, the coefficients of the:

second order expansion are	1, 2, 1
third order expansion are	1, 3, 3, 1
fourth order expasion are	1, 4, 6, 4, 1

In the thirteenth century the Chinese mathematician *Yang Hui* recognized that there was a predictable pattern in these numbers. By following his findings it is possible to predict that the next row of numbers is:

1, 5, 10, 10, 5, 1

which presumably must coincide with the coefficients of the fifth order binomial expansion. Indeed, it can easily be shown by multiplying the expansion of $(a + b)^4$ by $(a + b)$ that:

$$(a + b)^5 = a^5 + 5a^4b + 10a^3b^2 + 10a^2b^3 + 5ab^4 + b^5$$

so the prediction works. It was the French mathematician Blaise Pascal whose work using these numbers brought them to the attention of Western mathematicians some 400 years later in the seventeenth century. He constructed his now eponymous triangle:

```
            1
         1     1
      1     2     1
   1     3     3     1
 1     4     6     4     1
1    5    10    10    5    1
```

where a number in any position is obtained by adding the two numbers immediately above and directly to the left and right of that position. For example:

$1 + 2 = 3$ and $4 + 6 = 10$

Using Pascal's triangle of numbers we can generate the coefficients of any binomial expansion:

| **Binomial** | **Coefficients** |

$(a + b)^0$	1
$(a + b)^1$	1 1
$(a + b)^2$	1 2 1
$(a + b)^3$	1 3 3 1
$(a + b)^4$	1 4 6 4 1
$(a + b)^5$	1 5 10 10 5 1

Worked Examples

6.7 Derive the expansion of $(a+b)^4$ by multiplying the expansion of $(a+b)^3$ by $(a+b)$.

Solution:
$$(a+b)^4 = (a+b)(a+b)^3$$
$$= (a+b)(a^3 + 3a^2b + 3ab^2 + b^3)$$
$$= a(a^3 + 3a^2b + 3ab^2 + b^3) + b(a^3 + 3a^2b + 3ab^2 + b^3)$$
$$= a^4 + 3a^3b + 3a^2b^2 + ab^3 + ba^3 + 3a^2b^2 + 3ab^3 + b^4$$
$$= a^4 + 4a^3b + 6a^2b^2 + 4ab^3 + b^4$$

6.8 Extend Pascal's triangle to expand $(a+b)^6$.

Solution:

$(a+b)^5$		1		5		10		10		5		1	
$(a+b)^6$	1		6		15		20		15		6		1

Therefore:

$$(a+b)^6 \quad = a^6 + 6a^5b + 15a^4b^2 + 20a^3b^3 + 15a^2b^4 + 6ab^5 + b^6$$

6.9 Expand each of the following:
  (a) $(a-b)^3$ (b) $(2a+b)^5$
  (c) $(a+3b)^4$ (d) $(4a-5b)^3$
  (e) $(6a-b/6)^4$ (f) $(x+3)^6$

Solution:
(a) $(a-b)^3 = (a+[-b])^3$
$$= a^3 + 3a^2[-b] + 3a[-b]^2 + [-b]^3$$
$$= a^3 - 3a^2b + 3ab^2 - b^3$$

(b) $(2a+b)^5 = ([2a]+b)^5$
$$= [2a]^5 + 5[2a]^4b + 10[2a]^3b^2 + 10[2a]^2b^3 + 5[2a]b^4 + b^5$$
$$= 32a^5 + 80a^4b + 80a^3b^2 + 40a^2b^3 + 10ab^4 + b^5$$

(c) $(a+3b)^4 = (a+[3b])^4$
$$= a^4 + 4a^3[3b] + 6a^2[3b]^2 + 4a[3b]^3 + [3b]^4$$
$$= a^4 + 12a^3b + 54a^2b^2 + 108ab^3 + 81b^4$$

(d) $(4a-5b)^3 = ([4a]+[-5b])^3$
$$= [4a]^3 + 3[4a]^2[-5b] + 3[4a][-5b]^2 + [-5b]^3$$
$$= 64a^3 - 240a^2b + 300ab^2 - 125b^3$$

(e) $(6a-b/6)^4 = ([6a]+[-b/6])^4$

$$= [6a]^4 + 4[6a]^3[-b/6] + 6[6a]^2[-b/6]^2 + 4[6a][-b/6]^3 + [-b/6]^4$$
$$= 1296a^4 - 144a^3b + 6a^2b^2 - b/9 + b/1296$$

(f) $(x + 3)^6 = x^6 + 6x^5[3] + 15x^4[3]^2 + 20x^3[3]^3 + 15x^2[3]^4 + 6x[3]^5 + [3]^6$
$$= x^6 + 18x^5 + 135x^4 + 540x^3 + 1215x^2 + 1458x + 729$$

Exercises

6.7 Derive the expansion of $(a + b)^5$ by multiplying the expansion of $(a + b)^4$ by $(a + b)$.

6.8 Extend Pascal's triangle to expand $(a + b)^7$.

6.9 Expand each of the following:
  (a) $(a - b)^4$ (b) $(3a - b)^3$
  (c) $(a + b/5)^5$ (d) $(2a + 3b)^4$
  (e) $(3a + b/3)^3$ (f) $(x - 4)^5$

Unit 5   Fractions

Algebraic fractions

An algebraic fraction is in the form of a ratio of two algebraic expressions and can range from quite simple ratios such as:

$\dfrac{1}{a}$ or $\dfrac{x}{y}$

to more involved ratios such as:

$$\frac{ab^2 - cab}{3ba + 2ca^2}$$

Manipulating such fractions involves adding and subtracting them and cancelling out factors that are common to both the numerator and the denominator. For example, to add the fractions:

$$\frac{1}{a} + \frac{1}{b}$$

each fraction must be written as an equivalent fraction with a common denominator. In this example, we can write:

$$\frac{1}{a} = \frac{1}{a} \times \frac{b}{b} \quad \text{because} \quad \frac{b}{b} = 1$$

$$= \frac{b}{ab}$$

Similarly

$$\frac{1}{b} \times \frac{a}{a} = \frac{a}{ba} = \frac{a}{ab}$$

so that:

$$\frac{1}{a} + \frac{1}{b} = \frac{b}{ab} + \frac{a}{ab}$$

$$= \frac{b+a}{ab}$$

This process of adding two algebraic fractions by rewriting each as equivalent fractions with the same denominator is entirely the same procedure as that employed when adding two numeral fractions as was demonstrated in Part One.

Worked Examples

6.10   Perform the following additions and subtractions:

(a) $\dfrac{1}{a} - \dfrac{1}{b}$ (b) $\dfrac{b}{a} + \dfrac{a}{b}$

(c) $\dfrac{b}{a} + \dfrac{a}{c} - \dfrac{c}{b}$ (d) $\dfrac{x}{yz} + \dfrac{y}{xz} + \dfrac{z}{xy}$

(e) $\dfrac{3p}{4q} + \dfrac{2}{3q} - \dfrac{5q}{2p}$

Solution:

(a) $\dfrac{1}{a} - \dfrac{1}{b} = \dfrac{b}{ab} - \dfrac{a}{ab} = \dfrac{b-a}{ab}$

(b) $\dfrac{b}{a} + \dfrac{a}{b} = \dfrac{bb}{ab} + \dfrac{aa}{ba}$

$\quad = \dfrac{b^2}{ab} + \dfrac{a^2}{ab}$

$\quad = \dfrac{b^2 + a^2}{ab}$

(c) $\dfrac{b}{a} + \dfrac{a}{c} - \dfrac{c}{b} = \dfrac{bcb}{acb} + \dfrac{aab}{cab} - \dfrac{cac}{bac}$

$\quad = \dfrac{b^2 c}{abc} + \dfrac{a^2 b}{abc} - \dfrac{c^2 a}{abc}$

$\quad = \dfrac{b^2 c + a^2 b - c^2 a}{abc}$

(d) $\dfrac{x}{yz} + \dfrac{y}{xz} + \dfrac{z}{xy} = \dfrac{xx}{xyz} + \dfrac{yy}{xyz} + \dfrac{zz}{xyz}$

$\quad = \dfrac{x^2 + y^2 + z^2}{xyz}$

(e) $\dfrac{3p}{4q} + \dfrac{2}{3q} - \dfrac{5q}{2p} = \dfrac{3pp}{4qp} + \dfrac{2p}{3qp} - \dfrac{5qq}{2pq}$

$\quad = \dfrac{3p^2}{4qp} + \dfrac{2p}{3qp} - \dfrac{5q^2}{2pq}$

$\quad = \dfrac{9p^2}{12qp} + \dfrac{8p}{12qp} - \dfrac{30q^2}{12pq}$

$\quad = \dfrac{9p^2 + 8p - 30q^2}{12pq}$

Exercises

6.10 Perform the following additions and subtractions:

(a) $\dfrac{2}{a}+\dfrac{3}{b}$

(b) $\dfrac{x}{y}-\dfrac{y}{x}$

(c) $\dfrac{3}{p}-\dfrac{2}{q}-\dfrac{1}{pq}$

(d) $\dfrac{r}{st}-\dfrac{t}{rs}-\dfrac{s}{t}$

(e) $\dfrac{6a}{5b}-\dfrac{4}{5b}-\dfrac{2b}{3a}$

Cancelling common factors

In the following algebraic fraction:

$$\frac{ab+ac}{ap-aq}$$

both the numerator and the denominator can be factorized to give:

$$\frac{a(b+c)}{a(p-q)}$$

This reveals the fact that the numerator and the denominator have a factor in common and because:

$$\frac{a}{a}=1$$

we can say that:

$$\frac{a(b+c)}{a(p-q)}=\frac{(b+c)}{(p-q)}$$

thereby eliminating the common factor in both the numerator and the denominator. This procedure is known as **cancelling common factors**.

Worked Examples

6.11 Factorize the numerator and denominator in each of the following and cancel out any common factors:

(a) $\dfrac{x^2-xy}{xy-x}$

(b) $\dfrac{4a+6b}{2a-8b}$

(c) $\dfrac{p^2 - q^2}{p - q}$ (d) $\dfrac{u^2 - u - 12}{4 - u}$

(e) $\dfrac{a(a^2 + ab - 2b^2)}{a^2 - b^2}$

Solution:

(a) $\dfrac{x^2 - xy}{xy - x} = \dfrac{x(x - y)}{x(y - 1)}$

$\qquad = \dfrac{x - y}{y - 1}$

(b) $\dfrac{4a + 6b}{2a - 8b} = \dfrac{2(2a + 3b)}{2(a - 4b)}$

$\qquad = \dfrac{2a + 3b}{a - 4b}$

(c) $\dfrac{p^2 - q^2}{p - q} = \dfrac{(p - q)(p + q)}{p - q}$

$\qquad = p + q$

(d) $\dfrac{u^2 - u - 12}{4 - u} = \dfrac{(u + 3)(u - 4)}{4 - u}$

$\qquad = -\dfrac{(u + 3)(u - 4)}{u - 4}$

$\qquad = -(u + 3)$

(e) $\dfrac{a(a^2 + ab - 2b^2)}{a^2 - b^2} = \dfrac{a(a - b)(a + 2b)}{(a + b)(a - b)}$

$\qquad = \dfrac{a(a + 2b)}{(a + b)}$

6.12 Simplify each of the following:

(a) $\dfrac{xy - yz + 2yx - 2zy}{3(z - x)}$ (b) $\dfrac{p^2 q + q^2 p}{p(2q + 1) - q(p - 1) - (p + q)}$

(c) $\dfrac{a^2x+ay-x^2a-2ya}{x(3a-x)-2x^2-6y}$ (d) $\dfrac{a-1}{2ab-2b}+\dfrac{b+1}{2ab+2a}$

(e) $\left\{\dfrac{a/b-b/a}{abc^2}\right\}\left\{\dfrac{abc}{a+b}\right\}$

Solution:

(a) $\dfrac{xy-yz+2yx-2zy}{3(z-x)}=\dfrac{3xy-3yz}{3(z-x)}$

$\qquad\qquad =\dfrac{3y(x-z)}{3(z-x)}$

$\qquad\qquad =\dfrac{-3y(z-x)}{3(z-x)}$

$\qquad\qquad =-y$

(b) $\dfrac{p^2q+q^2p}{p(2q+1)-q(p-1)-(p+q)}=\dfrac{pq(p+q)}{2pq+p-qp+q-p-q}$

$\qquad\qquad\qquad =\dfrac{pq(p+q)}{pq}$

$\qquad\qquad\qquad =p+q$

(c) $\dfrac{a^2x+ay-x^2a-2ya}{x(3a-x)-2x^2-6y}=\dfrac{ax(a-x)-2ay}{3ax-x^2-2x^2-6y}$

$\qquad\qquad\qquad =\dfrac{ax(a-x)-2ay}{3ax-3x^2-6y}$

$\qquad\qquad\qquad =\dfrac{ax(a-x)-2ay}{3x(a-x)-6y}$

$\qquad\qquad\qquad =\dfrac{a[x(a-x)-2y]}{3[x(a-x)-2y]}$

$\qquad\qquad\qquad =\dfrac{a}{3}$

(d) $\dfrac{a-1}{2ab-2b}+\dfrac{b+1}{2ab+2a}=\dfrac{a-1}{2b(a-1)}+\dfrac{b+1}{2a(b+1)}$

$\qquad\qquad\qquad =\dfrac{1}{2b}+\dfrac{1}{2a}$

$$= \frac{a}{2ab} + \frac{b}{2ab}$$

$$= \frac{a+b}{2ab}$$

(e) $\left\{ \dfrac{a/b - b/a}{abc^2} \right\} \left\{ \dfrac{abc}{a+b} \right\} = \left\{ \dfrac{a}{ab^2c^2} - \dfrac{b}{a^2bc^2} \right\} \left\{ \dfrac{abc}{a+b} \right\}$

$$= \left\{ \frac{a^2}{a^2b^2c^2} - \frac{b^2}{a^2b^2c^2} \right\} \left\{ \frac{abc}{a+b} \right\}$$

$$= \left\{ \frac{a^2 - b^2}{a^2b^2c^2} \right\} \left\{ \frac{abc}{a+b} \right\}$$

$$= \left\{ \frac{(a-b)(a+b)}{abc} \right\} \left\{ \frac{1}{a+b} \right\}$$

$$= \left\{ \frac{a-b}{abc} \right\}$$

Exercises

6.11 Factorize the numerator and denominator in each of the following and cancel out any common factors:

(a) $\dfrac{pq - p}{qp - 2p}$

(b) $\dfrac{rs + st - s}{rst + s}$

(c) $\dfrac{3f - 15g}{6f + 3g}$

(d) $\dfrac{7x + 14y - 21z}{14y - 7z}$

(e) $\dfrac{u^2 - v^2}{u + v}$

6.12 Simplify each of the following:

(a) $\dfrac{3ab + bc - 2ba - 2cb}{b(c - a)}$

(b) $\dfrac{2xy - 2x + y - 1}{4(1 - y)}$

(c) $\dfrac{u^2v - v^2u}{2u^2v^3 - 2v^2u^3}$

(d) $\dfrac{a^3 - b^3}{a(a + 2b) + b(b - 2a)}$

(e) $\dfrac{s^2r + ts^2 - r^2s - rst}{sr - s}$

Module 6   Further exercises

1   Combine like terms in each of the following algebraic expressions:
(a) $3pq - 5qr - 4rp + 2rq - qp$
(b) $-3r - 4rs - s + 8sr$
(c) $11uvw - 9wvu + 2vwu - 3vw - uw$
(d) $13x^3y + 5x - 4yx^3 + 3x - 8y + 9y^3x$
(e) $-3ab^2 + 4b^5c - 2a^2c - 11abc^4 + 7cb^{-1} - 6ab^5 + 9a^2c - 12c^4ba$

2   Factorize similar terms in each of the following algebraic expressions:
(a) $-3xy + 11yz$ (b) $7ab + ac - 2db - 3dc$
(c) $5rst - 2st - rs + 3s$ (d) $6p^2qr - 4pr - 5pq + 4$

3   Simplify each of the following algebraic expressions:
(a) $8ba - 7cb + 2ab - bc$
(b) $-8w - 5wv - 2u + vu$
(c) $-15xyz - 4zxy + 3yzx + 8xy - zy$
(d) $15p^2r^2 - 10p^2s - 24qr^3 + 16qs$
(e) $2ca^2 - 2a^2b - 2abc + 2ab^2$

4   Expand each of the following algebraic expressions:
(a) $q(7 - 2p)$ (b) $2l(3m - 7n)$
(c) $(6x + 4y)3z$ (d) $4s(2r - 5t) - 2s(5r - 2t)$
(e) $(3u + 4w)2v - (4v - 3w)2u$

5   Multiply out each of the following algebraic expressions:
(a) $(3a + b)(c - 3d)$ (b) $(p - 5q)(7p + 3q)$
(c) $(x^2 + 5)(2x^2 + 3x - 4)$ (d) $(2x - 3)(5x^3 - 4x^2 + 3x - 2)$
(e) $(8 - 3x - 4x^2)(2x^3 - 5x^2 + 3x + 4)$

6   Factorize each of the following algebraic expressions:
(a) $x^2 + 8x + 15$ (b) $x^2 + 2x - 15$
(c) $x^2 + x - 20$ (d) $x^2 + 8x + 16$
(e) $x^2 - 81$ (f) $4x^2 - 36$
(g) $x^3 - 8$ (h) $8x^3 + 27$
(j) $54x^3 - 16$ (k) $24x^3 + 192$

7   Derive the expansion of $(a + b)^6$ by multiplying the expansion of $(a + b)^5$ by $(a + b)$.

8   Extend Pascal's triangle to expand $(a + b)^8$.

9   Expand each of the following:
   (a) $(a-b)^3$ (b) $(5a-b)^4$
   (c) $(a-5b)^4$ (d) $(4a+6b)^3$
   (e) $(a+1/a)^4$ (f) $(a/b-b/a)^3$

10 Perform the following additions and subtractions:

   (a) $\dfrac{1}{x}-\dfrac{1}{xy}$ (b) $\dfrac{2}{a}+\dfrac{a}{2}$

   (c) $\dfrac{3f}{g}+\dfrac{2g}{f}-\dfrac{4}{5}$ (d) $\dfrac{6}{abc}-\dfrac{5}{ab}+\dfrac{1}{c}$

   (e) $\dfrac{7u}{3v}-\dfrac{3v}{2w}+\dfrac{9w}{5u}$

11 Factorize the numerator and denominator in each of the following and cancel out any common factors:

   (a) $\dfrac{n^3-m^3}{n^2+nm+m^2}$ (b) $\dfrac{x^2-3x+2}{x-1}$

   (c) $\dfrac{x^3-5x^2+6x}{x^2-2x}$ (d) $\dfrac{n^3-nm^2}{n^2-nm}$

   (e) $\dfrac{3x^2-9x+6}{6x^2+12x-18}$

12 Simplify each of the following:

   (a) $\dfrac{x(x^2-y^2)+y(y^2-xy)}{y^2-x(y-x)}$ (b) $\dfrac{p-q}{pq^2}-\dfrac{p-q}{p^2q}$

   (c) $\dfrac{1}{1/u+1/v}$ (d) $\dfrac{m^2/n+n^2/m}{m+n}$

   (e) $\dfrac{a^3-b^3+(a-b)ab}{a^3-ab^2+ba^2-b^3}$

Module 7

Expressions, equations and graphs

OBJECTIVES

When you have completed this module you will be able to:

- ■ Numerically evaluate an algebraic expression

- ■ Construct an algebraic equation from an algebraic expression, distinguishing between independent and dependent variables

- ■ Construct the graph of an algebraic equation involving a one-variable algebraic expression

- ■ Convert a discrete graph into a continuous graph and perform interpolation and extrapolation procedures

- ■ Derive and use the equation of a straight line

There are four units in this module:

Unit 1: Algebraic equations
Unit 2: Graphs of algebraic equations
Unit 3: Graphs and information
Unit 4: The equation of a straight line

Unit 1   Algebraic equations

Try the following test:

1   Construct an appropriate equation for each of the following expressions and tabulate the values of the dependent variable corresponding to each of the given values of the independent variable:
   (a) $3x + 4$; $x = 0, 2, 4, 6, 8$
   (b) $2x^2 - 3x$; $x = -3, -1, 0, 1, 3$
   (c) $2x^2 + 10x + 12$; $x = -4, -2, 0, 2, 4$
   (d) $3x + 3/x$; $x = 1, 2, 3, 4, 5$
   (e) $x^3 - 2x^2 - x + 2$; $x = -2, -1, 0, 1, 2$

Variables and their values

In unit 2 of module 5 we defined a variable to be a placeholder for any one of a range of numbers and in everything that we have considered since then we have tacitly assumed that the range of numbers referred to is the entire collection of real numbers. However, there may be occasions when we wish to restrict the range of numbers to be a smaller collection.

Imagine, if you will, that you run a small business from your home selling gift packs consisting of six flavours of jam. After deducting the production, packing and distribution costs you find that each pack that you sell makes you £3.00 profit so that if you sell x packs in a week your gross profit for the week is:

$3x$ pounds

However, every week you place an advertisement in the local paper to advertise your wares and this costs you £4.00 each week. Consequently, the total net profit in pounds that you make on your enterprise by selling x packs of jam in a week is:

$3x - 4$ pounds

Notice that in this expression we have restricted the numbers that x represents to the whole numbers because you only sell packs of jam in whole packs.

Using this expression for the net profit we can easily see that if you sell 10 packs in a week you make:

$3 \times 10 - 4 = 26$ pounds net profit

and if you sell 15 packs in a week you make:

$3 \times 15 - 4 = 41$ pounds net profit

By **assigning** values to the variable x in this way we obtain corresponding values for the expression as a whole. This process is referred to as **numerically evaluating** the expression. By assigning a **range** of values to the variable x we can obtain a corresponding range of values for the expression as a whole. This means that we can use a variable as a placeholder for the range of values taken on by the expression – we can represent the entire expression by a single variable. If we use the letter y as the variable to represent the value of the expression for the net profit we can then write:

$$y = 3x - 4$$

where y represents the profit in pounds obtained by selling x gift packs of jam.

We have formed an **equation** involving the two variables x and y where the value of the variable y depends upon the value assigned to the variable x. When we numerically evaluate the expression we select at will any one of the permitted numbers that x represents and so find the corresponding value of the variable y. Accordingly, we call x the **independent variable** because it is chosen at will and we call y the **dependent variable**; the value of y depends upon the independently selected value of x.

Worked Examples

7.1 Construct an appropriate equation for each of the following expressions and tabulate the values of the dependent variable corresponding to each of the given values of the independent variable:

(a) $5x - 6$; $x = 0, 1, 2, 3$
(b) $x^2 + x$; $x = -1, 0, 1$
(c) $3x^2 - 2x + 4$; $x = -2, 0, 2, 3$
(d) $7x + 6/x$; $x = 1, 3, 6, 12$
(e) $x^3 - 6x^2 + 11x - 6$; $x = -2, -1, 0, 1, 2, 3$

Solution:
(a) $y = 5x - 6$ where $x = 0, 1, 2, 3$

x :	0	1	2	3
$5x$:	0	5	10	15
-6 :	-6	-6	-6	-6
y :	-6	-1	4	9

(b) $y = x^2 + x$ where $x = -1, 0, 1$

x :	-1	0	1
x^2 :	1	0	1
x :	-1	0	1
y :	0	0	2

(c) $y = 3x^2 - 2x + 4$ where $x = -2, 0, 2, 3$

x :	-2	0	2	3
$3x^2$:	12	0	12	27
$-2x$:	4	0	-4	-6
4 :	4	4	4	4
y :	20	4	12	25

(d) $y = 7x + 6/x$ where $x = 1, 3, 6, 12$

x :	1	3	6	12
$7x$:	7	21	42	84
$6/x$:	6	2	1	0.5
y :	13	23	43	84.5

(e) $y = x^3 - 6x^2 + 11x - 6$ where $x = -2, -1, 0, 1, 2, 3$

x :	-2	-1	0	1	2	3
x^3 :	-8	-1	0	1	8	27
$-6x^2$:	-24	-6	0	-6	-24	-54
$11x$:	-22	-11	0	11	22	33
-6 :	-6	-6	-6	-6	-6	-6
y :	-60	-24	-6	0	0	0

Exercises

7.1 Construct an appropriate equation for each of the following expressions and tabulate the values of the dependent variable corresponding to each of the given values of the independent variable:
(a) $8 - 3x$; $x = 1, 3, 5, 7$
(b) $2x^2 - x$; $x = -2, 0, 2$
(c) $x^2 + 4x + 4$; $x = -3, -2, -1, 0$
(d) $x - 1/x$; $x = 2, 4, 8, 16$
(e) $x^3 + 3x^2 + 3x + 1$; $x = -2, -1, 0, 1, 2$

Unit 2   Graphs of algebraic equations

Try the following test:

1   Write down the co-ordinates of each of the labelled points in the following Figure:

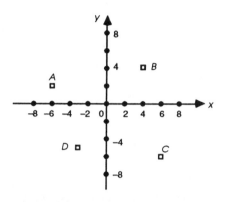

2   In each of the following the corresponding values of two variables are tabulated. From each table construct a collection of ordered pairs and then plot them against a cartesian co-ordinate system (the first variable values in each case are to be drawn on the horizontal co-ordinate axis):

(a)
x:	0	1	2	3
y:	0	6	12	18

(b)
p:	2	4	6	8
q:	1	7	7	1

(c)
u:	−2	−1	0	1	2
v:	2	0	−2	0	2

(d)
s:	−2	−1	0	1	2
t:	3	0	−1	0	3

An equation and its graph

The French mathematician *René Descartes*, who was born in 1596, was responsible for devising an ingenious method of linking algebraic equations to geometric shapes. An apocryphal tale is told of how he lay in bed late one morning, as was his habit, watching a fly walk in a straight line across his bedroom ceiling. As he lay there it dawned on him that the path traced out by the fly could be described by an algebraic equation if only one knew the positions of the fly relative to the two adjacent walls. True or not, the tale contains within it the essence of what *Descartes* achieved.

Let us assume that a square grid consisting of lines parallel to the adjacent South and East walls was marked out on *Descartes'* bedroom ceiling. Further, let us assume that, on four different occasions, *Descartes* read off from the grid the positions of the fly against the South and East walls of his bedroom as follows:

$$S: \quad 1 \qquad 2 \qquad 3 \qquad 4$$
$$E: \quad 2 \qquad 3 \qquad 4 \qquad 5$$

where the numbers refer to grid units. Rising from his bed *Descartes* then drew the following picture where the points on the grid show the location of the fly on each of the four occasions:

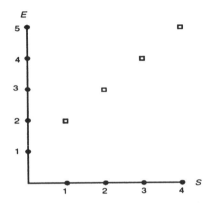

The letters S and E refer to the South and East walls of the room respectively.

Over a rather late breakfast *Descartes* looked at this picture and realized that he could link the four measurements in a single equation. Using S and E as variables to represent the respective positions relative to the South and West walls it is a simple matter to link the positions in the following equation:

$$E = S + 1$$

Just check. When $S = 1$, then:

$E = 1 + 1$ by using the equation

$\quad = 2$ Correct as checked from the table of numbers

When $S = 3$, then:

$E = 3 + 1$ by using the equation
$\quad = 4$ Correct as checked from the table of numbers

How about positions he hadn't measured? What if at one time the fly's position against the S wall was 2.5, what would be its position against the E wall?

Use the equation:

$$E = 2.5 + 1 = 3.5.$$

How do we know that this prediction is correct?

Looking closely at the picture we see that all four points can be joined together by a straight line – the path of the fly's tour across the ceiling. If we do do this we obtain the following Figure:

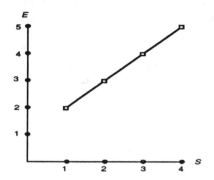

Now, if we mark off 2.5 on the *S* wall and draw a line parallel to the grid up to the straight line and then at right angles to the East wall we read off 3.5. Magic! – it works!

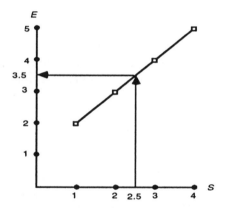

The equation:

$$E = S + 1$$

not only describes the four measurements taken by *Descartes* but also describes all those measurements that he could have taken had he been bothered to do so. The fly had walked across the ceiling in a straight line and the equation that we have just derived describes that fact **in algebraic form**. We call the equation the **equation of this straight line** and the picture we call the **graph** of this straight line – we have linked an algebraic equation to a geometric shape.

Co-ordinate axes and plotted points

The construction of the graph that we have just drawn was achieved against a framework of two, mutually perpendicular, straight lines on each of which were marked a collection of numbered points. This framework is called the **cartesian co-**

ordinate system in honour of *Descartes* and the two, mutually perpendicular straight lines are called the **co-ordinate axes**. The co-ordinate axes are, in fact two real lines (straight lines on which are plotted the real numbers) that intersect in their common zero point – we call this point the **origin** of the co-ordinate system.

The cartesian co-ordinate system lies in a plane and every point in the plane can be uniquely referenced against the co-ordinate system. For example, in the following Figure the co-ordinate system axes are labelled *x* and *y* respectively The point in the plane labelled *P* is referenced by *x* and *y* values obtained by reading off the respective co-ordinate axes the numbers immediately below and immediately to the left of the point *P*.

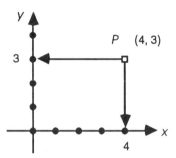

These two numbers are called the **co-ordinates** of point *P* and are listed within brackets where, by convention, the first number refers to the horizontal co-ordinate and the second, the vertical co-ordinate. We call these two numbers within the bracket an **ordered** pair of numbers because their order is important. That is, for example:

(1, 2) and (2, 1)

are two different ordered pairs and represent two different points in the plane.

Just as we can reference every point in the plane by an ordered pair of numbers set against a cartesian co-ordinate system so we can go the other way. That is, given an ordered pair of numbers we can find the point in the plane to which it refers.

For example, in the following Figure we find that the ordered pair (3, 2) refers to point *Q*.

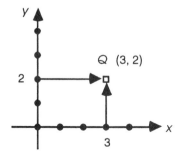

Locating the point *Q* in this way is a process known as **plotting** the point (3, 2).

Worked Examples

7.2 Write down the co-ordinates of each of the labelled points in the following Figure:

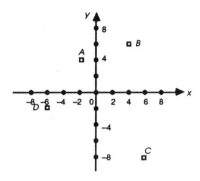

Solution:
The co-ordinates are:

$$A \quad (-2, 4)$$
$$B \quad (4, 6)$$
$$C \quad (6, -8)$$
$$D \quad (-6, -2)$$

7.3 In each of the following the corresponding values of two variables are tabulated. From each table construct a collection of ordered pairs and then plot them against a cartesian co-ordinate system (the first variable values in each case are to be drawn on the horizontal co-ordinate axis:

(a) x: 0   1   2   3
    y: −2 −1   0   1

(b) p: 2   4   6   8
    q: 6   4   2   0

(c) u: −2 −1   0   1   2
    v: 0   3   6   9 12

(d) s: −2 −1   0   1   2
    t: 2   1   0 −1 −2

Solution:
(a) The ordered pairs are:

(0, −2), (1, −1), (2, 0), (3, 1). The graph is as follows:

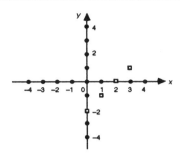

(b) The ordered pairs are:

   $(2, 6), (4, 4), (6, 2), (8, 0)$. The graph is as follows:

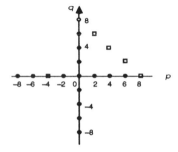

(c) The ordered pairs are:

   $(-2, 0), (-1, 3), (0, 6), (1, 9), (2, 12)$

(d) The ordered pairs are:

   $(-2, 2), (-1, 1), (0, 0), (1, -1), (2, -2)$

Exercises

7.2 Write down the co-ordinates of each of the labelled points in the following Figure:

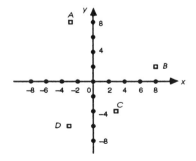

7.3 In each of the following the corresponding values of two variables are tabulated. From each table construct a collection of ordered pairs and then plot them against a cartesian co-ordinate system (the first variable values in each case are to be drawn on the horizontal co-ordinate axis:

(a) x: 0 1 2 3
    y: 4 2 0 −2

(b) p: 2 4 6 8
    q: 6 10 14 18

(c) u: −2 −1 0 1 2
    v: 4 1 0 1 4

(d) s: −2 −1 0 1 2
    t: −8 −1 0 1 8

Unit 3　Graphs and information

Try the following test:

1　Plot each of the following complete tabulated lists of data and in each
graph join the points together with a straight line that extends beyond the
plotted data. Use the line to find the missing values of the variables in the
second, incomplete, table, distinguishing between the processes of
interpolation and extrapolation:

(a)
x:	0	1	2	3
y:	−4	−1	2	5

x:	0.5	4	?	?
y:	?	?	6	0.5

(b)
p:	2	4	6	8
q:	9	13	17	21

p:	0	?	10
q:	?	5	?

(c)
u:	−2	−1	0	1	2
v:	4	1	−2	−5	−8

u:	?	0.5	?
v:	−10	?	5

Discrete and continuous graphs

All the graphs that we have plotted so far have consisted of isolated points and all
graphs of this nature are referred to as **discrete** graphs. When *Descartes* joined up
the points in his discrete graph to obtain the straight line he converted his discrete
graph into a **continuous** graph. However, he could do this only because he knew
beforehand that all the points that he plotted lay on a straight line – the fly walked
across his ceiling in a straight line. If he had not known this beforehand then joining
up the points in a straight line would have been pure guesswork. We shall return to
the problem of converting discrete graphs into continuous graphs shortly.

　　All discrete graphs have a problem in common – it is often difficult to see a
collection of isolated plotted points. One of the reasons for drawing a graph in the
first place is to give a pictorial representation to a set of data so illegibility tends to
defeat the purpose. To overcome this problem discrete graphs are often highlighted
either by joining the points together with a jagged line or, preferably, by constructing
vertical or horizontal bars from a co-ordinate axis to each point.

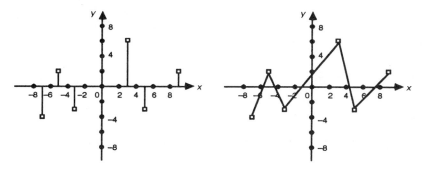

It must always be remembered that the highlighting bars are not a part of the plotted graph but are there merely as an aid to the eye. One of the problems associated with a jagged line graph is that the jagged line tends to impart to the viewer more information than is really there.

Graphs as a store of information

Whilst *Lewis Carroll* hunts the *Snark*, the mathematician hunts the pattern, for within the pattern lies the secret of understanding; the commonality of disparate ideas is often revealed when each idea displays the same pattern. This is one of the reasons why we use graphs to display numeric data – to see if there is any recognizable pattern in the data. In the examples and exercises that you have just looked at the data was stored in a tabulated list prior to it being plotted on a cartesian graph. In list form any pattern in the data was not immediately obvious but once plotted it was clearly seen that every point in each set lay in a straight line. This is the type of information that is stored in a graph; a visual representation of the pattern in data. From such patterns general assumptions can be made which, if correct, lead to even more detailed information being gleaned from the graph.

When *Descartes* joined the four isolated points of his discrete graph with a straight line to form a continuous graph he was **assuming** that if he plotted every single point of the fly's path he would have obtained the continuous straight line. This assumption was, of course a correct one to make because he had seen the fly walk in a straight line. The assumption also overcomes a fundamental problem associated with each and every continuous graph – **they are all impossible to plot**. It is impossible to plot a continuous graph because it would require an infinite amount of time to plot an infinite number of points. Joining the finite number of plotted points is the best we can do and we should always be aware of this, especially if we are going to use the continuity so gained to glean additional information.

Once we have joined the discrete points together in this way the graph now contains an abundance of information that it did not possess hitherto. For example, we are told that a car's fuel supply depends upon the distance it has travelled and an experiment is performed to test this statement. The following is the result:

Kilometres travelled	:	0	9	18	27	36
Litres remaining	:	30.0	28.5	27.0	25.5	24.0

If this data is plotted against a cartesian co-ordinate system we obtain the following graph:

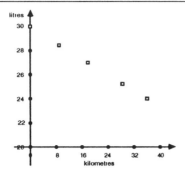

From the plotted points it seems a reasonable assumption that we can join the points with a straight line which we do in the following graph:

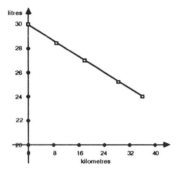

Interpolation and extrapolation
Having joined the points with a straight line we are now able to obtain values of kilometres travelled and litres remaining that are not contained within the list of tabulated values. For example, we see that for 4.5 kilometres travelled there are 29.25 litres remaining:

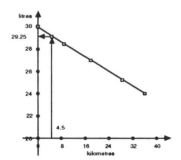

Gleaning information that is contained within the collection of points by finding the value of the dependent variable from an un-listed value of the independent variable is a process referred to as **interpolation**.

We can also interpolate the other way. For example, if we have 26.25 litres remaining then we have travelled 22.5 kilometres:

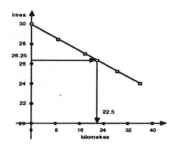

Interpolating from the dependent variable to the independent variable in this way is referred to as **inverse interpolation**.

We can also make the additional assumption that the relationship between kilometres travelled and litres remaining will continue to be correct beyond the listed data. This assumption will enable us to extend the line beyond the points plotted:

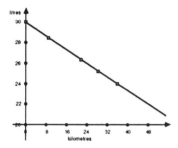

Now we can see that for 45 kilometres travelled there are 22.5 litres remaining:

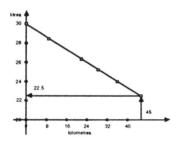

Gleaning information that is contained beyond the collection of points in this way is a process referred to as **extrapolation**. Care must be taken when extrapolating data because it may not always be a reasonable procedure to perform. For example, if we measure the temperature in a greenhouse at hourly intervals between 7:00 am and midday on a sunny summer morning we may find that a regular increase in temperature would suggest a linear relationship between time and temperature simply because the sun was shining all morning. For this reason, interpolating between measured values will probably be a reasonable procedure to perform. However, extrapolating beyond the recorded data assumes that the weather conditions remained the same after the data gathering ceased which may or may not be true.

Worked Examples

7.4 Plot each of the following complete tabulated lists of data and in each graph join the points together with a straight line that extends beyond the plotted data. Use the line to find the missing values of the variables in the second, incomplete, table, distinguishing between the processes of interpolation and extrapolation:

(a)

x:	0	1	2	3
y:	−2	−1	0	1

x:	0.5.	4	?	?
y:	?	?	−3	0.5

(b)

p:	2	4	6	8
q:	6	4	2	0

p:	0	?	10
q:	?	5	?

(c)

u:	−2	−1	0	1	2
v:	0	3	6	9	12

u:	?	0.5	?
v:	−1	?	15

Solution:
(a) The graph is as follows:

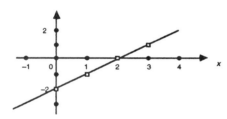

From the graph we can complete the table as follows:

x:	0.5	4	−1	2.5	*extrapolation : interpolation*
y:	−1.5	2	−3	0.5	*interpolation : extrapolation*

(b) From the graph we can complete the table as follows:

p:	0	3	10	*interpolation*
q:	8	5	−2	*extrapolation : extrapolation*

(c) From the graph we can complete the table as follows:

u:	$-7/3$	0.5	3	extrapolation : extrapolation
v:	-1	15/2	15	interpolation

Exercises

7.4 Plot each of the following complete tabulated lists of data and in each graph join the points together with a straight line that extends beyond the plotted data. Use the line to find the missing values of the variables in the second, incomplete, table, distinguishing between the processes of interpolation and extrapolation:

(a)
x:	0	1	2	3
y:	-4	-2	0	2

x:	0.5	4	?	?
y:	?	?	-3	0.5

(b)
p:	2	4	6	8
q:	-6	-4	-2	0

p:	0	?	10
q:	?	5	?

(c)
u:	-2	-1	0	1	2
v:	12	9	6	3	0

u:	?	0.5	?
v:	-1	?	15

Unit 4   The equation of a straight line

The straight line

We began this chapter by creating and evaluating an algebraic equation and, thereby, obtaining a collection of ordered pairs of numbers. These ordered pairs were then plotted to give a graph of the equation – from the equation we plotted the graph. Can we go the other way? From the graph can we derive the equation? We have seen a lot of straight lines recently; can we derive their equations from their graphs? To answer these questions it is best to start by looking at a straight line and finding out what is common for all straight lines and what is intrinsic to a specific line.

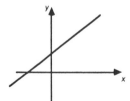

■ *Every straight line crosses either one or both of the co-ordinate axes*:
    A vertical line crosses just the *x*-axis and a horizontal line crosses just the
    y-axis. Any other line crosses both axes.

■ *Every straight line possesses a slope*:
    Every line makes an angle with the horizontal axis.

■ *Only one straight line can be drawn through two specified points*: Any two given points will only lie on one specific line.

Consider the line in the following Figure which crosses the y-axis at $y = b$:

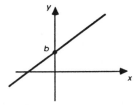

We can see that the point at which the line crosses the y-axis is given by the ordered pair:

$(0, b)$

Next, we construct a right-angled triangle whose hypotenuse coincides with the line by selecting a point on the line with co-ordinates:

(x, y)

and drawing the vertical from that point to meet the horizontal line from $(0, b)$ to meet at:

(x, b)

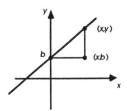

The slope of the line in this Figure is measured by the amount the line rises in the y direction as the line is traversed in the direction of increasing x. We can say that, for a straight line, the ratio:

$$\frac{\text{rise in the } y \text{ direction}}{\text{travel in the } x \text{ direction}}$$

is a constant; we shall signify this constant by the letter a. This means that if the line is traversed from where it intersects the y-axis to the point (x, y), then:

$$\frac{y - b}{x} = a$$

We now wish to rewrite this equation by isolating the variable y on the left-hand side. To do this we perform arithmetic operations on the equation making sure that whatever operation we perform on the left-hand side we also, simultaneously, perform on the right-hand side so as to maintain the balance of the equation. The first action we take is to multiply both sides of the equation by x as this will eliminate the variable x on the left-hand side of the equation:

$$\frac{(y-b)x}{x} = ax$$

That is:

$$y - b = ax$$

Now, add b to both sides of this equation so as to eliminate b from the left-hand side of the equation and thereby isolating the variable y:

$$y - b + b = ax + b$$

Giving:

$$y = ax + b$$

This is the **equation of the straight line** with slope a that intercepts the y-axis at the point $y = b$. We talk about the **family** of straight lines where every member of the family has an equation of this form. For example, the equation:

$$y = 2x + 3$$

is the equation of the straight line with slope 2 and vertical intercept 3.

Notice that any member of the family of straight lines whose equations are of the form:

$$y = 2x + b$$

has the same gradient as the line:

$$y = 2x + 3$$

but a different vertical intercept (if $b \neq 3$). As a consequence, all such lines are parallel:

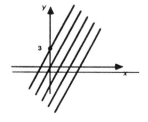

Also, any member of the family of straight lines whose equations are of the form:

$$y = ax + 3$$

has the same vertical intercept as the line:

$$y = 2x + 3$$

but a different gradient (if $a \neq 2$).

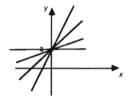

Variables and constants
For any given member of the family of straight lines:

$$y = ax + b$$

the values of a and b are **fixed** but the values of x and y are **variable**, the latter being the co-ordinates of any point that lies on the line. Accordingly, whilst symbols a and b are variables for a family of straight lines they are referred to as **constants** of the equation because they are constant for any particular member of the family. Also, whilst it is not a hard and fast rule, we do tend to use letters near the beginning of the alphabet for constants and letters near the end of the alphabet for variables so as to give a visual distinction between the two types of symbol.

Worked Examples

7.5 Plot each of the following straight lines over the stated ranges:
   (a) $y = 5x - 2 : -2 \leq x \leq 2$
   (b) $y = -2x + 3 : -1 \leq x \leq 2$
   (c) $p = 5 - 10q : -0.5 \leq q \leq 1$
   (d) $s = 2t + 3 : -2 \leq t \leq 1$

Solution:
(a) $y = 5x - 2 : -2 \leq x \leq 2$

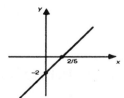

(b) $y = -2x + 3 : -1 \leq x \leq 2$

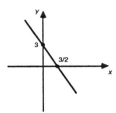

(c) $p = 5 - 10q : -0.5 \leq q \leq 1$

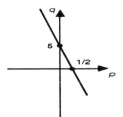

(d) $s = 2t + 3 : -2 \leq t \leq 1$

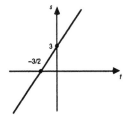

7.6 In each of the following, find the equation of the straight line that satisfies the stated conditions:

(a) Has gradient 5 and vertical intercept -4

(b) Has gradient -2 and contains the point $(3, -4)$

(c) Contains the point $(5, 0)$ and has vertical intercept 15

(d) Contains the pair of points $(3, 1)$ and $(5, 7)$

(e) Is parallel to the line:

$$y = -2x + 9$$

and passes through the point $(5, -3)$

Solution:

(a) Has gradient 5 and vertical intercept -4

The equation of the line is of the form $y = ax + b$. In this particular case $a = 5$ and $b = -4$. The equation is then:

$y = 5x - 4$

(b) Has gradient –2 and contains the point (3, –4)

The equation of the line is of the form $y = ax + b$. In this particular case $a = -2$ so the equation is of the form:

$y = -2x + b$

We are given that (3, –4) lies on the line, so substituting this pair of co-ordinates into the above equation we find that:

$-4 = (-2)3 + b$

That is:

$-4 = -6 + b$

So that $b = 2$ and the equation of the line is:

$y = -2x + 2$

(c) Contains the point (5, 0) and has vertical intercept 15

The equation of the line is of the form $y = ax + b$. In this particular case $b = 15$ so the equation is of the form:

$y = ax + 15$

We are given that (5, 0) lies on the line, so substituting this pair of co-ordinates into the above equation we find that:

$0 = a(5) + 15$

That is:

$0 = 5a + 15$ so that $a = -(1/3)$ and the equation of the line is:

$y = (-1/3)x + 15$

(d) Contains the pair of points (3, 1) and (5, 7)

The equation of the line is of the form $y = ax + b$. We are given that (3, 1) and (5, 7) both lie on the line, so substituting these two pairs of co-ordinates into the above equation we find that:

$1 = 3a + b$
$7 = 5a + b$

That is:

$7 = 5a + b$
$\quad = 3a + b + 2a$

but:

$3a + b = 1$

giving:

$7 = 1 + 2a$

So that $2a = 6$ and hence, $a = 3$. Substituting this back into the first equation we find that:

$1 = 9 + b$

So that $b = -8$ and the equation of the line is:

$y = 3x - 8$

(e) Any line parallel to the line:

$y = -2x + 9$

is given by:

$y = 2x + b$

The line passes through the point $(5, -3)$ so:

$-3 = 10 + b$

So that $b = -13$ and the equation of the line is:

$y = 2x - 13$

Exercises

7.5 Plot each of the following straight lines over the stated ranges:
(a) $y = 6x - 3 : -1 \leq x \leq 1.5$
(b) $y = -3x + 2 : -0.5 \leq x \leq 1.5$

    (c) $v = 4 - 12u : -0.5 \leq u \leq 0.5$
    (d) $m = 4n + 4 : -1.5 \leq n \leq 0.5$

7.6 In each of the following, find the equation of the straight line that satisfies the stated conditions:

    (a) Has gradient 7 and vertical intercept -3
    (b) Has gradient -5 and contains the point $(-2, -1)$
    (c) Contains the point $(8, 0)$ and has vertical intercept 4
    (d) Contains the pair of points $(1, 5)$ and $(5, 1)$
    (e) Is parallel to the line:

$$y = -5x - 2$$

    and passes through the point $(-4, 2)$

Module 7　Further exercises

1　Construct an appropriate equation for each of the following expressions and tabulate the values of the dependent variable corresponding to each of the given values of the independent variable:
  (a) $2x - 3$;　　　　　　　　$x = 1, 3, 5, 7$
  (b) $2x - 3x^2$;　　　　　　　$x = 0, 1, 2$
  (c) $x^2 + x - 2$;　　　　　　$x = -2, -1, 0, 1, 2$
  (d) $2x + 1/4x$;　　　　　　　$x = 2, 4, 8, 16$
  (e) $x^3 + 6x^2 + 11x + 6$;　　$x = -5, -4, -3, -2, -1, 0$

2　Write down the co-ordinates of each of the labelled points in the following Figure:

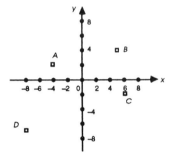

3　In each of the following the corresponding values of two variables are tabulated. From each table construct a collection of ordered pairs and then plot them against a cartesian co-ordinate system (the first variable values in each case are to be drawn on the horizontal co-ordinate axis:

  (a) x:　　0　　1　　2　　3　　　　　(b) p:　　2　　4　　6　　8
  　　y:　　-2　-1　-1　-2　　　　　　　q:　　8　　6　　4　　2

  (c) u:　-2　-1　　0　　1　　2　　　(d) s:　-2　-1　　0　　1　　2
  　　v:　-4　-2　　0　　2　　4　　　　　t:　　2　　1　　0　　1　　2

4　Plot each of the following complete tabulated lists of data and in each graph join the points together with a straight line that extends beyond the plotted data. Use the line to find the missing values of the variables in the second, incomplete, table, distinguishing between the processes of interpolation and extrapolation:

  (a) x:　　0　　1　　2　　3　　　　　(b) p:　　2　　4　　6　　8
  　　y:　　11　　8　　5　　2　　　　　　q:　　13　31　49　67

$x:$	0.5	4	?	?
$y:$	?	?	6	0.5

$p:$	0	?	10
$q:$	?	6	?

(c)

$u:$	−2	−1	0	1	2
$v:$	0	−2	−4	−6	−8

$u:$	?	0.5	?
$v:$	−1	?	15

5 Plot each of the following straight lines over the stated ranges:
   (a) $y = 3x − 7 : −1 \le x \le 1.5$
   (b) $y = −4x − 2 : −0.5 \le x \le 1.5$
   (c) $v = −13 − 11u : −0.5 \le u \le 0.5$
   (d) $m = 2n + 1 : −1.5 \le n \le 0.5$

6 In each of the following, find the equation of the straight line that satisfies the stated conditions:
   (a) Has gradient −2 and vertical intercept 5
   (b) Has gradient 11 and contains the point (9, −9)
   (c) Contains the point (−6, 1) and has vertical intercept 2
   (d) Contains the pair of points (3, 9) and (−6, 4)
   (e) Is parallel to the line:

   $$y = −x − 1$$

   and passes through the point (−1, 1)

Module 8

Manipulating equations

OBJECTIVES

When you have completed this module you will be able to:

- Transpose the dependent and independent variables in an equation

- Convert an equation in two variables to either one of two alternative forms

- Describe the order and component terms of a polynomial

- Solve linear and quadratic equations

- Determine the number of real solutions to a quadratic equation from a study of its discriminant.

There are four units in this module:

Unit 1: Changing the subject
Unit 2: Polynomials: expressions and equations
Unit 3: Solving quadratic equations
Unit 4: Quadratic equations in general

Unit 1 Changing the subject

Try the following test:

1 Transpose each of the following equations:
   (a) $y = 3x - 4$ (b) $u = 2v^2 - 7$
   (c) $m = (2n - 1)^3 - 3$ (d) $a = [16 - (2b^{3/4})]^{1/3}$
   (e) $p = [(q^2 + 4)^{3/2} - 9]^{1/3}$

2 From each of the following equations obtain two alternative forms
   where each alternative form has a dependent variable isolated on the
   left-hand side:
   (a) $3x - 4y = 9$ (b) $x^2 + y^2 = 5$
   (c) $(u/3)^2 + (v/4)^2 = 2$ (d) $(f/2)^2 - (g/5)^2 = 1$

3 Find the range of values of the independent variable for which the
   dependent variable has real values in each of the following:
   (a) $x^2 + y^2 = 5$ (b) $(u/3)^2 + (v/4)^2 = 2$
   (c) $(f/2)^2 - (g/5)^2 = 1$

Transposition of variables

An algebraic equation represents a balancing of numerical values on either side of
the equation expressed in symbolic form. As a consequence, any arithmetic operations performed on one side of the equation must be duplicated on the other side so
as to maintain the balance of the equation.

Many typical equations will consist of the **subject** of the equation as the
dependent variable on the left-hand side by itself and the independent variable
embedded within some expression on the right-hand side. For example:

$$p = (3q^2 - 5)^{1/3}$$

Here, p is the dependent variable and the subject of the equation whilst q is the
independent variable. By performing a sequence of arithmetical operations simultaneously on both sides of this equation, the variables p and q can be **transposed** to
form a different equation where q is the dependent variable and subject of the new
equation and p is the independent variable. To effect the transposition of p and q we
must adopt a strategy.

To find the numerical value of the dependent variable of an equation for a
selected value of the independent variable we perform a sequence of arithmetic
operations. To effect the transposition of the two variables we must **undo each one
of these arithmetic operations in turn**, starting with the last one of the sequence and
working through to the first of the sequence. We undo an arithmetic operation by
performing its **inverse** operation. For example, consider the evaluation of p from a

given value of q in the equation:

$$p = (3q^2 - 5)^{1/3}$$

The secret is to imagine that you are using a calculator and, starting with a given value for q, you are going to find the corresponding value for p.

Enter a value for q

Operation
raise to the power 2
multiply by 3
subtract 5
raise to the power 1/3

This will give the corresponding value of p.

Next, alongside each operation in this list write down the operation that reverses the effect – the inverse operations, that is:

Operation	**Inverse operation**
raise to the power 2	raise to the power 1/2
multiply by 3	divide by 3
subtract 5	add 5
raise to the power 1/3	raise to the power 3

The sequence of operations for transposition is then given by the list of inverse operations in reverse order, namely:

raise to the power 3
add 5
divide by 3
raise to the power 1/2

Try it:

$$p = (3q^2 - 5)^{1/3}$$

Raise both sides of the equation to the power 3:

$$\begin{aligned} p^3 &= [(3q^2 - 5)^{1/3}]^3 \\ &= (3q^2 - 5)^1 \\ &= 3q^2 - 5 \end{aligned}$$

Add 5 to both sides of the equation:

$$\begin{aligned} p^3 + 5 &= 3q^2 - 5 + 5 \\ &= 3q^2 \end{aligned}$$

Divide both sides of the equation by 3:

$$(p^3 + 5)/3 = (3q^2)/3$$
$$= q^2$$

Raise both sides of the equation to the power 1/2:

$$[(p^3 + 5)/3]^{1/2} = [q^2]^{1/2}$$
$$= q$$

Finally, to put the equation into the normal form of subject on the left and expression on the right, we interchange sides of the equation to produce the final transposed result:

$$q = [(p^3 + 5)/3]^{1/2}$$

Here, as you can see, we now have an equation where the two variables have been transposed in that p is now the independent variable and q is the dependent variable and subject of the new equation.

Independent and dependent variables
In the previous module we introduced the concept of an equation of the form:

$y =$ some expression involving a variable x

In such an equation it is clear that we evaluate y for a suitably selected value of x; the variable y depends upon the value chosen for the independently chosen value of the variable x. There are, however, equations where the distinction between the dependent and the independent variable is not immediately obvious, indeed, the choice of which is which may be quite arbitrary. Take for example the circle in the following Figure:

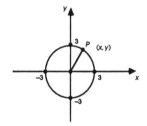

Here the circle of radius 3 has its centre at the co-ordinate origin. If we select any point P on the circumference of the circle and denote its location by the ordered pair (x, y) then we can use *Pythagoras'* theorem to deduce the **equation of the circle**. We shall merely state the theorem here leaving a proof of the theorem to the next section. *Pythagoras'* theorem states that, for a right-angled triangle ABC:

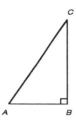

The square of the hypotenuse AC is equal to the sum of the squares of the other two sides AB and BC

In terms of the lengths of the sides of the triangle this means that

$$AB^2 + BC^2 = AC^2$$

Relating this to the circle and the point P on its circumference we see that this translates to:

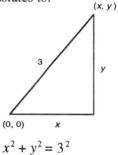

$$x^2 + y^2 = 3^2$$

That is:

$$x^2 + y^2 = 9$$

and this is the equation of the circle centred on the co-ordinate origin with radius 3 – the co-ordinates of any point on the circle satisfy this equation. The equation of the general circle centred on the origin and having radius r is:

$$x^2 + y^2 = r^2$$

and if a circle has its centre at the point (a, b) its equation is:

$$(x - a)^2 + (y - b)^2 = r^2$$

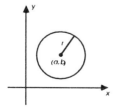

Here the distinction between the independent and the dependent variable is quite arbitrary and we can evaluate such an equation by first selecting a value for **either** *x* or *y*. When plotting the graphs of such equations the only requirement that we demand is that whatever the choice of independent variable it is always plotted on the horizontal axis.

The straight line
An alternative form for the equation of a straight line that permits an arbitrary choice of variable type is:

$px + qy = r$

where the values of *p. q* and *r* are constant for a specific straight line. For example, the equation:

$2x + 3y = 6$

can be plotted as follows, choosing the *x*-axis to be the horizontal axis:

Let $x = 0$, giving:

$0 + 3y = 6$, that is $y = 2$ so that the point $(0, 2)$

lies on the line. Now let $y = 0$, giving:

$2x + 0 = 6$, that is $x = 3$ so that the point $(3, 0)$

lies on the line. We can now plot these two points and then join them to produce the graph of the equation:

$2x + 3y = 6$

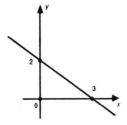

Notice that the equation with the number 12 replacing the number 6:

$2x + 3y = 12$

can be plotted by joining the points $(0, 4)$ and $(6, 0)$

to produce a line which is parallel to the first one. Indeed, if we use the straight line equation:

$$px + qy = r$$

then different r-values will correspond to different members of a family of straight lines, all of whose members are parallel.

To demonstrate the equivalence between this form of the equation for a straight line and the one given previously, namely:

$$y = ax + b \text{ and } px + qy = r$$

we transpose the second form to make the variable y the subject of the equation.

Given:

$$px + qy = r$$

subtract px from both sides of the equation to yield:

$$px + qy - px = r - px, \text{ that is:}$$

$$qy = r - px$$
$$\quad = -px + r$$

Now, divide both sides of the equation by q to yield:

$$y = (-px + r)/q$$
$$\quad = (-p/q)x + (r/q)$$

This is now in the original form of the equation of the straight line:

$$y = ax + b$$

where the gradient is:

$$a = (-p/q)$$

and the vertical intercept is:

$$b = (r/q)$$

Worked Examples

8.1 Transpose each of the following equations:
    (a) $y = 5x - 6$ (b) $s = 3t^2 + 4$
    (c) $u = (5v - 2)^3 + 7$ (d) $p = [8 - (4q^{4/5})]^{1/2}$
    (e) $m = [(n^2 + 3)^{1/3} - 4]^5$

Solution:

(a) $y = 5x - 6$

Add 6:

$$y + 6 = 5x - 6 + 6$$
$$= 5x$$

Divide by 5:

$$(y + 6)/5 = (5x)/5$$
$$= x$$

Interchange sides:

$$x = (y + 6)/5$$
$$= y/5 + 6/5$$

(b) $s = 3t^2 + 4$

Subtract 4:

$$s - 4 = 3t^2 + 4 - 4$$
$$= 3t^2$$

Divide by 3:

$$(s - 4)/3 = (3t^2)/3$$
$$= t^2$$

Raise to power 1/2:

$$[(s - 4)/3]^{1/2} = [t^2]^{1/2}$$
$$= t$$

Interchange sides:

$$t = [(s - 4)/3]^{1/2}$$

(c) $u = (5v - 2)^3 + 7$

Subtract 7:

$$u - 7 = (5v - 2)^3 + 7 - 7$$
$$= (5v - 2)^3$$

Raise to power 1/3:

$$(u - 7)^{1/3} = [(5v - 2)^3]^{1/3}$$
$$= 5v - 2$$

Add 2:

$$(u - 7)^{1/3} + 2 = 5v - 2 + 2$$
$$= 5v$$

Divide by 5:

$$[(u - 7)^{1/3} + 2]/5 = [5v]/5$$
$$= v$$

Interchange sides:

$$v = [(u - 7)^{1/3} + 2]/5$$

(d) $p = [8 - (4q^{4/5})]^{1/2}$

Raise to power 2:

$$p^2 = 8 - (4q^{4/5})$$

Subtract 8:

$$p^2 - 8 = -4q^{4/5}$$

Divide by -4:

$$(p^2 - 8)/(-4) = q^{4/5}$$

That is:

$$(8 - p^2)/4 = q^{4/5}$$

Raise to power 5/4:

$$[(8 - p^2)/4]^{5/4} = q$$

Interchange sides:

$$q = [(8 - p^2)/4]^{5/4}$$

(e) $m = [(n^2 + 3)^{1/3} - 4]^5$

Raise to power 1/5 and then add 4:

$$m^{1/5} + 4 = (n^2 + 3)^{1/3}$$

Raise to power 3 and then subtract 3:

$$[m^{1/5} + 4]^3 - 3 = n^2$$

Raise to power 1/2 and then interchange sides:

$$n = ([m^{1/5} + 4]^3 - 3)^{1/2}$$

8.2 From each of the following equations obtain two alternative forms where each alternative form has a dependent variable isolated on the left-hand side:

(a) $5x - 7y = 35$ (b) $x^2 + y^2 = 16$

(c) $(p/5)^2 + (q/2)^2 = 4$ (d) $(u/3)^2 - (v/4)^2 = 1$

Solution:

(a) $5x - 7y = 35$

$$5x = 35 + 7y$$
$$x = 7 + (7/5)y \text{ or } x = (7/5)y + 7$$

or:

$$-7y = 35 - 5x$$
$$y = -5 + (5/7)x \text{ or } y = (5/7)x - 5$$

(b) $x^2 + y^2 = 16$

$$x^2 = 16 - y^2$$
$$x = (16 - y^2)^{1/2}$$

or:

$$y = (16 - x^2)^{1/2}$$

(c) $(p/5)^2 + (q/2)^2 = 4$

$$(p/5)^2 = 4 - (q/2)^2$$
$$p/5 = [4 - (q/2)^2]^{1/2}$$
$$p = 5[4 - (q/2)^2]^{1/2}$$

or:

$$(q/2)^2 = 4 - (p/5)^2$$
$$q/2 = [4 - (p/5)^2]^{1/2}$$
$$q = 2[4 - (p/5)^2]^{1/2}$$

(d) $(u/3)^2 - (v/4)^2 = 1$

$$(u/3)^2 = 1 + (v/4)^2$$

$$u/3 = [1 + (v/4)^2]^{1/2}$$
$$u = 3[1 + (v/4)^2]^{1/2}$$

or:

$$-(v/4)^2 = 1 - (u/3)^2$$
$$(v/4)^2 = (u/3)^2 - 1$$
$$v/4 = [(u/3)^2 - 1]^{1/2}$$
$$v = 4[(u/3)^2 - 1]^{1/2}$$

8.3 Find the range of values of the independent variable for which the dependent variable has real values in each of the following:
(a) $x^2 + y^2 = 16$ (b) $(p/5)^2 + (q/2)^2 = 4$
(c) $(u/3)^2 - (v/4)^2 = 1$

Solution:
(a) From the equation $x^2 + y^2 = 16$ in question 8.2(b) above, we obtained the two alternative forms:

$$x = (16 - y^2)^{1/2} \text{ or } y = (16 - x^2)^{1/2}$$

The first form requires that for x to be a real number, $16 - y^2 \geq 0$ (the square roots of negative numbers do not produce real numbers). We transpose across inequalities in just the same manner as transposing across equalities. That is, given:

$$16 - y^2 \geq 0$$

add y^2 to both sides to yield

$$16 \geq y^2 \text{ or, by interchanging sides, } y^2 \leq 16$$

To find the corresponding inequality for y requires taking the square root. However, the square root of 16 produces the two possible values ± 4. This means that:

$$y \leq 4 \text{ or } y \geq -4$$

We can put both of these two statements into one, thus:

$$-4 \leq y \leq 4$$

The second form requires that for y to be real, $16 - x^2 \geq 0$. That is:

$$-4 \leq x \leq 4$$

(b) From the equation $(p/5)^2 + (q/2)^2 = 4$ in question 8.2(c) above, we obtained the two alternative forms:

$$p = 5[4 - (q/2)^2]^{1/2} \text{ or } q = 2[4 - (p/5)^2]^{1/2}$$

The first form requires that for p to be real, $4 - (q/2)^2 \geq 0$. That is:

$$-4 \leq q \leq 4$$

The second form requires that for q to be real, $4 - (p/5)^2 \geq 0$. That is:

$$-10 \leq p \leq 10$$

(c) From the equation $(u/3)^2 - (v/4)^2 = 1$ in question 8.2(d) above, we obtained the two alternative forms:

$$u = 3[1 + (v/4)^2]^{1/2} \text{ or } v = 4[(u/3)^2 - 1]^{1/2}$$

The first form requires that for u to be real, $1 + (v/4)^2 \geq 0$ which it is for all real values of v.

The second form requires that for v to be real, $(u/3)^2 - 1 \geq 0$. That is:

$$-\sqrt{3} \leq u \leq +\sqrt{3}$$

Exercises

8.1 Transpose each of the following equations:
(a) $y = -7x + 2$
(b) $a = 8 - 5b^2$
(c) $h = 2 - (8k + 3)^4$
(d) $u = [(7v^{2/3}) + 5]^{1/3}$
(e) $w = [13 + (z^3 - 9)^{2/5}]^5$

8.2 From each of the following equations obtain two alternative forms where each alternative form has a dependent variable isolated on the left-hand side:
(a) $8y + 6x = 24$
(b) $x^2 + y^2 = 49$
(c) $(m/2)^2 + (n/7)^2 = 16$
(d) $(a/6)^2 - (b/3)^2 = 1$

8.3 Find the range of values of the independent variable for which the dependent variable has real values in each of the following:
(a) $x^2 + y^2 = 49$
(b) $(a/6)^2 - (b/3)^2 = 1$
(c) $(m/2)^2 + (n/7)^2 = 16$

Unit 2   Polynomial expressions and equations

Try the following test:

1   In each of the following, write the polynomial in ascending or descending order, give the order of the polynomial, and list the coefficients in descending power order:
   (a) $18 - 2x^2 - x^3 + 6x$
   (b) $16x + x^2 - x^3 - 10$
   (c) $2x^4 - 1$
   (d) $3 - 12x^5$
   (e) $1 - x^8$

2   Solve each of the following linear equations:
   (a) $y = 3x - 7$ when $y = 11$
   (b) $y = 2 + 4x$ when $y = -5$
   (c) $4x - y = -5$ when $y = -1$
   (d) $6x - 7 = 0$
   (e) $9 + 3x = 0$

3   Use transposition and substitution to solve each of the following pairs of simultaneous equations:
   (a) $5x - 3y = -19$
       $x + 4y = 10$
   (b) $3x + 2y = -3$
       $y = 3x - 6$

   (c) $9x^2 - 16y^2 = 12$
       $3x + 4y = -2$
   (d) $3y^2 + 7xy + 2x^2 = 25$
       $y + 2x = 5$

Polynomial expressions in a single variable
An algebraic expression in the form of a sum or difference of terms where each term contains a single variable raised to a natural number power is called a **polynomial expression in a single variable**. For example, the algebraic expression:

$$3x^4 - 5x^3 + 0.3x^2 + x - 0.4$$

is a **fourth order polynomial expression** in the variable x – fourth order because the highest power is 4. The terms of a polynomial expression are written in ascending or descending order of powers and each term contains within it a multiplying factor called the **coefficient** of the term – notice that the last term, the so-called **constant term**, could have been written as:

$0.4x^0$ because $x^0 = 1$

so that 0.4 is the coefficient of the constant term.

An expression of the form:

$$ax + b$$

where a and b are the constant coefficients is a polynomial expression of degree 1 – also referred to as a **linear expression** because of its relationship to a straight line. The polynomial expression of degree 2, also known as a **quadratic expression**, is of the form:

$$ax^2 + bx + c$$

where a, b and c are the constant coefficients. The polynomial expression of degree three, also known as the **cubic expression**, is of the form:

$$ax^3 + bx^2 + cx + d$$

where a, b, c and d are the constant coefficients.

Polynomial equations

Any equation of the form:

$$y = \text{some polynomial expression}$$

is referred to as a **polynomial equation**. The polynomial equation of degree 1:

$$y = ax + b$$

is called a **linear equation** and the polynomial equation of degree 2:

$$y = ax^2 + bx + c$$

is called a **quadratic equation**.

Worked Examples

8.4 In each of the following, write the polynomial in ascending or descending order, give the order of the polynomial, and list the coefficients in descending power order:
   (a) $7x - 2x^2 + x^3 - 8$
   (b) $-3x + 2$
   (c) $2x^5 - 3x^3 + x - 5$
   (d) $x - 9x^4 + 2x^3 + 4x^2 + 5$
   (e) $5.4x^6 - 9.3$

Solution:
(a) $7x - 2x^2 + x^3 - 8$

   $x^3 - 2x^2 + 7x - 8$, Order 3, Coefficients 1, –2, 7, –8 respectively

(b) $-3x + 2$, Order 1, Coefficients -3, 1 respectively

(c) $2x^5 - 3x^3 + x - 5$

   $2x^5 + 0x^4 - 3x^3 + 0x^2 + x - 5$, Order 5, Coefficients 2, 0, -3, 0, 1, -5 respectively

(d) $x - 9x^4 + 2x^3 + 4x^2 + 5$

   $5 + x + 4x^2 + 2x^3 - 9x^4$, Order 4, Coefficients -9, 2, 4, 1, 5 respectively

(e) $5.4x^6 - 9.3$

   $5.4x^6 + 0x^5 + 0x^4 + 0x^3 + 0x^2 + 0x - 9.3$, Order 6,

   Coefficients 5.4, 0, 0, 0, 0, 0, -9.3 respectively

Exercises

8.4 In each of the following, write the polynomial in ascending or descending order, give the order of the polynomial, and list the coefficients in descending power order:

(a) $-2x + 5x^2 - 1$ (b) $3x - x^2 + 4$

(c) $7x - 9$ (d) $18 - 3x$

(e) $7x^3 - x + 2$ (f) $18 - 2x^2 - x^3 + 6$

(g) $16x + x^2 - x^3 - 10$ (h) $2x^4 - 1$

(j) $3 - 12x^5$ (k) $1 - x^8$

Solving linear equations

When we evaluate an equation in two variables we derive the value of the dependent variable for a given value of the independent variable. For example, given the linear equation:

$$y = 3x - 4$$

we easily see that when $x = 5$ then:

$$y = 3 \times 5 - 4$$
$$= 11$$

When we **solve** an equation we reverse this process; we look for the value or values of the independent variable for a given value of the dependent variable. For example, given the equation:

$$y = 3x - 4$$

what is the value of x when $y = 14$? To answer this question we first substitute the numerical value for y to yield the equation:

$$14 = 3x - 4$$

Solving this equation now entails **finding that value of x which satisfies it.** To do this clearly requires a familiarity with the processes of transposition; we add 4 to both sides:

$$14 + 4 = 3x$$

and then divide both sides by 3 to give the final result:

$$18/3 = x$$

That is:

$$x = 6$$

Worked Examples

8.5 Solve each of the following linear equations:
    (a) $y = 2x - 4$ when $y = 6$ (b) $y = 6 + 5x$ when $y = 10$
    (c) $6x - 3y = 18$ when $y = 3$ (d) $5x - 3 = 0$
    (e) $3 - 2x = 0$

Solution:
(a) $y = 2x - 4$ when $y = 6$

    When $y = 6$ then:

$$6 = 2x - 4$$

    That is:

$$2x = 10, \text{ so } x = 5$$

(b) $y = 6 + 5x$ when $y = 10$

    When $y = 10$ then:

$$10 = 6 + 5x$$

    That is:

$$5x = 4, \text{ so } x = 4/5$$

(c) $6x - 3y = 18$ when $y = 3$

When $y = 3$ then:

$6x - 9 = 18$

That is:

$6x = 27$, so $x = 9/2$

(d) $5x - 3 = 0$

Since $5x - 3 = 0$:

$5x = 3$

That is:

$x = 3/5$

(e) $3 - 2x = 0$

Since $3 - 2x = 0$:

$2x = 3$

That is:

$x = 3/2$

Exercises

8.5 Solve each of the following linear equations:
  (a) $y = 5x - 6$ when $y = 14$ (b) $y = 3 + 7x$ when $y = 4$
  (c) $5x - 2y = 10$ when $y = 4$ (d) $8x - 5 = 0$
  (e) $9 - 7x = 0$

Solving simultaneous equations
When we first encountered the concept of a variable as a placeholder for a number we considered the problem of determining the prices charged for apples and bananas after two consecutive purchases. To refresh your memory, on day 1, 2 apples and 4 bananas were purchased for £1 and on day 2, 4 apples and 2 bananas were purchased for £1.40. Putting this information into algebraic form we have:

$2A + 4B = 100$ and $4A + 2B = 140$

Each of these two equations is an equation of a straight line. If we plot these two equations on the same graph you will see that the lines cross at the point with co-ordinates:

$A = 30, B = 10$

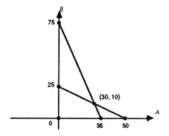

Each line consists of an infinity of points but only one point is simultaneously on both lines and this one point represents the simultaneous solution to the problem. To solve the problem algebraically we can employ one of two methods. We can use the method used in module 5 or, alternatively, we can use a method that employs transposition and substitution:

Transpose one of the equations to find one variable in terms of the other. For example, given:

$2A + 4B = 100$

we can transpose this to write:

$A = 50 - 2B$

This can now be substituted into the second equation to give:

$4(50 - 2B) + 2B = 140$

That is:

$200 - 6B = 140$

so that:

$B = 10$ and hence:

$A = 50 - 2B$
$ = 30$

The required solution.

Worked Examples

8.6 Use transposition and substitution to solve each of the following pairs of simultaneous equations:

(a) $2x + 3y = 16$
$5x - 2y = 2$

(b) $4x + 3y = 1$
$y = 2x + 7$

(c) $x^2 - y^2 = 4$
$x + y = 2$

(d) $2x^2 - 4y^2 = -2$
$x - 2y = 1$

Solution:
(a) $2x + 3y = 16$
$5x - 2y = 2$

From the first equation we see that:

$x = 8 - (3/2)y$

Substituting this into the second equation yields:

$5(8 - (3/2)y) - 2y = 2$

That is:

$40 - (19/2)y = 2$

so:

$y = 4$ and hence $x = 2$

(b) $4x + 3y = 1$
$y = 2x + 7$

From the second equation we see that:

$y = 2x + 7$

Substituting this into the first equation yields:

$4x + 3(2x + 7) = 1$

That is:

$$10x = -20$$

so:

$$x = -2 \text{ and hence } y = 3$$

(c) $x^2 - y^2 = 4$
$x + y = 2$

Factorizing the first equation we find that:

$$(x + y)(x - y) = 4$$

The second equation tells us that $x + y = 2$, consequently:

$$2(x - y) = 4 \text{ or } x - y = 2$$

We now have two linear equations to solve, namely:

$$x + y = 2$$
$$x - y = 2$$

This has the solution:

$$x = 2 \text{ and } y = 0$$

(d) $2x^2 - 4y^2 = -2$
$x - 2y = 1$

From the second equation we see that:

$$x = 2y + 1$$

Substituting into the first equation we obtain:

$$2(2y + 1)^2 - 4y^2 = -2$$

that is:

$$8y^2 + 8y + 2 - 4y^2 = -2$$

so that:

$$4y^2 + 8y + 4 = 0$$

That is:

$$4(y + 1)^2 = 0$$

The required solution is then:

$$y = -1 \text{ and } x = -1$$

Exercises

8.6 Use transposition and substitution to solve each of the following pairs of simultaneous equations:

(a) $4x + 5y = 31$
    $3x - 2y = 6$

(b) $7x + 5y = 2$
    $y = 2x - 3$

(c) $4x^2 - 9y^2 = 6$
    $2x - 3y = 3$

(d) $2y^2 + 4y - 8x^2 = 4$
    $y - 3x = -1$

Unit 3   Solving quadratic equations

Try the following test:

1   Solve each of the following quadratic equations:
  (a) $y = 1 - x - x^2$ when $y = -5$ (b) $y = x^2 + 6x + 16$ when $y = 8$
  (c) $y = 16x^2 - 8x + 1$ when $y = 9$ (d) $y = 9x^2 - 16$ when $y = -7$
  (e) $x^2 - 4 = 0$

2   Solve each of the following quadratic equations by completing the
  squares:
  (a) $x^2 + 3x - 1 = 0$ (b) $x^2 - 7x + 7 = 0$
  (c) $x^2 - 5x - 11 = 0$ (d) $5x^2 + x - 1 = 0$
  (e) $2 - 6x + 4x^2 = 0$

Solving quadratic equations

The general quadratic equation is of the form:

$$y = ax^2 + bx + r$$

where x and y are variables and a, b and r are constants for any given quadratic.

If we assign the value s to the variable y we obtain the equation:

$$s = ax^2 + bx + r$$

or, by subtracting s from both sides of the equation:

$$0 = ax^2 + bx + (r - s)$$

Because the constants r and s combine in this way to produce another constant, call it c, we see that solving **any** quadratic equation is tantamount to solving the equation:

$$0 = ax^2 + bx + c$$

For example, consider the quadratic equation:

$$y = x^2 - 3x + 6$$

To solve this equation when $y = 4$ we see that we must solve the equation:

$$4 = x^2 - 3x + 6$$

That is, we must solve the equation:

$$x^2 - 3x + 2 = 0$$

Noting that we can use standard factorization on the algebraic expression on the left-hand side of this equation we see that we can rewrite this equation as:

$$(x - 1)(x - 2) = 0$$

This means that either:

$$x - 1 = 0 \text{ or } x - 2 = 0$$

In turn this means that:

$$\text{either } x = 1 \text{ or } x = 2$$

which is the required solution. These two values are also referred to as the **roots** of the quadratic equation. This is a reasonable word to use because if we are given the two values:

$$x = 1 \text{ or } x = 2$$

we can say that:

$$x - 1 = 0 \text{ or } x - 2 = 0$$

Consequently:

$$(x - 1)(x - 2) = 0$$

That is:

$$x^2 - 3x + 2 = 0$$

the quadratic can be generated directly from its roots.

Notice that if we plot the graph of the equation:

$$y = x^2 - 3x + 2$$

we obtain the a shape called a **parabola**:

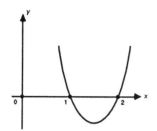

The values of x where the graph crosses the x-axis are those points where:

$$x^2 - 3x + 2 = 0$$

and as such represent the solutions to the quadratic equation.

Worked Examples

8.7 Solve each of the following quadratic equations:
    (a) $y = x^2 + x + 5$ when $y = 7$ (b) $y = 3x^2 - 4x - 12$ when $y = -8$
    (c) $x^2 - x - 6 = 0$ (d) $x^2 - 10x + 25 = 0$
    (e) $4x^2 - 49 = 0$

Solution:
(a) $y = x^2 + x + 5$ when $y = 7$

Substituting 7 for y in this equation gives:

$$7 = x^2 + x + 5$$

Subtracting 7 from both sides yields:

$$0 = x^2 + x - 2$$

The quadratic expression on the left-hand side factorizes to give:

$$0 = (x - 1)(x + 2)$$

This means that:

$$(x - 1) = 0 \text{ or } (x + 2) = 0$$

That is:

$x = 1$ or $x = -2$ and this is the required solution.

(b) $y = 3x^2 - 4x - 12$ when $y = -8$

Substituting -8 for y in this equation gives:

$$-8 = 3x^2 - 4x - 12$$

Adding 8 to both sides yields:

$$0 = 3x^2 - 4x - 4$$

The quadratic expression on the left-hand side factorizes to give:

$$0 = (3x + 2)(x - 2)$$

This means that:

$$(3x + 2) = 0 \text{ or } (x - 2) = 0$$

That is:

$x = -2/3$ or $x = 2$ and this is the required solution.

(c) $x^2 - x - 6 = 0$

The quadratic expression on the left-hand side factorizes to give:

$$(x + 2)(x - 3) = 0$$

This means that:

$$(x + 2) = 0 \text{ or } (x - 3) = 0$$

That is:

$x = -2$ or $x = 3$ and this is the required solution.

(d) $x^2 - 10x + 25 = 0$

The quadratic expression on the left-hand side factorizes to give:

$$(x - 5)(x - 5) = (x - 5)^2 = 0$$

A perfect square. This means that:

$$(x - 5) = 0$$

That is:

$x = 5$ and this is the required solution.

(e) $4x^2 - 49 = 0$

The quadratic expression on the left-hand side is a difference of two squares and so factorizes to give:

$(2x - 7)(2x + 7) = 0$

This means that:

$(2x - 7) = 0$ or $(2x + 7) = 0$

That is:

$x = \pm 7/2$ and this is the required solution.

Exercises

8.7 Solve each of the following quadratic equations:
(a) $y = x^2 - 2x + 3$ when $y = 6$ (b) $y = 9 - x - x^2$ when $y = 7$
(c) $y = 4x^2 + 4x - 8$ when $y = -5$ (d) $y = 6x^2 - 13x - 2$ when $y = -8$
(e) $x^2 - x - 12 = 0$

Completing the squares
In all the problems of the previous exercises the resultant quadratic that was required to be solved had a standard factorization. However, not all quadratic equations are immediately amenable to this method of solution. For example, the quadratic expression on the left-hand of the quadratic equation:

$x^2 + 6x + 2 = 0$

cannot be factorized using standard factors. To solve this particular problem we make use of a procedure known as **completing the squares**. The first two terms of the quadratic are the same as the first two terms of a quadratic that is a perfect square:

$x^2 + 6x$ are the same as the first two terms of the expansion of $(x + 3)^2$

Indeed, because:

$(x + 3)^2 = x^2 + 6x + 9$

we can write:

$x^2 + 6x = (x + 3)^2 - 9$

so that:

$$x^2 + 6x + 2 = (x + 3)^2 - 9 + 2$$
$$= (x + 3)^2 - 7$$

The quadratic equation:

$$x^2 + 6x + 2 = 0$$

can now be written as:

$$(x + 3)^2 - 7 = 0$$

That is:

$$(x + 3)^2 = 7$$

so that, taking square roots of both sides:

$$x + 3 = \pm\sqrt{7}$$

and hence:

$$x = \pm\sqrt{7} - 3$$

These are the two solutions to the quadratic equation, namely:

$$x = -3 + \sqrt{7} \text{ or } x = -3 - \sqrt{7}$$

Worked Examples

8.8 Solve each of the following quadratic equations by completing the squares:
(a) $x^2 + 4x + 1 = 0$ (b) $x^2 - 6x + 3 = 0$
(c) $x^2 + 3x - 3 = 0$ (d) $4x^2 - 12x + 5 = 0$
(e) $2 - 4x - 4x^2 = 0$

Solution:
(a) $x^2 + 4x + 1 = 0$

$$(x + 2)^2 = x^2 + 4x + 4 \text{ so that:}$$

$$x^2 + 4x + 1 = (x + 2)^2 - 4 + 1$$
$$= (x + 2)^2 - 3$$

The equation can then be written as:

$$(x + 2)^2 - 3 = 0 \ or \ (x + 2)^2 = 3$$

so that $x + 2 = \pm\sqrt{3}$

therefore:

$$x = \pm\sqrt{3} - 2$$

(b) $x^2 - 6x + 3 = 0$

$(x - 3)^2 = x^2 - 6x + 9$ so that:

$$x^2 - 6x + 3 = (x - 3)^2 - 9 + 3$$
$$= (x - 3)^2 - 6$$

The equation can then be written as:

$$(x - 3)^2 = 6$$

so that $x - 3 = \pm\sqrt{6}$

therefore:

$$x = 3 \pm \sqrt{6}$$

(c) $x^2 + 3x - 3 = 0$

$(x + 3/2)^2 = x^2 + 3x + 9/4$ so that:

$$x^2 + 3x - 3 = (x + 3/2)^2 - 9/4 - 3$$
$$= (x + 3/2)^2 - 21/4$$

The equation can then be written as:

$$(x + 3/2)^2 = 21/4$$

so that $x + 3/2 = \pm(\sqrt{21})/2$

therefore:

$$x = -3/2 \pm (\sqrt{21})/2$$

(d) $4x^2 - 12x + 5 = 0$

$(2x - 3)^2 = 4x^2 - 12x + 9$ so that:

$$4x^2 - 12x + 5 = (2x - 3)^2 - 9 + 5$$
$$= (2x - 3)^2 - 4$$

The equation can then be written as:

$$(2x - 3)^2 = 4$$

so that $2x - 3 = \pm 2$

therefore:

$$x = 3/2 \pm 1$$

(e) $2 - 4x - 4x^2 = 0$

$$-(1 + 2x)^2 = -1 - 4x - 4x^2 \text{ so that:}$$

$$2 - 4x - 4x^2 = -(1 + 2x)^2 + 1 + 2$$
$$= -(1 + 2x)^2 + 3$$

The equation:

$$2 - 4x - 4x^2 = 0$$

can then be written as:

$$(1 + 2x)^2 = 3$$

so that $1 + 2x = \pm\sqrt{3}$

therefore:

$$x = -1/2 \pm (\sqrt{3})/2$$

Exercises

8.8 Solve each of the following quadratic equations by completing the squares:
(a) $x^2 + 8x + 5 = 0$ (b) $x^2 - 10x + 15 = 0$
(c) $x^2 + 5x + 5/4 = 0$ (d) $9x^2 - 6x - 3 = 0$
(e) $1 + 4x - x^2 = 0$

Unit 4   Quadratic equations in general

The general quadratic equation

Solving quadratic equations as we have been doing has necessitated using one of two different methods – the choice of method being dependent upon the particular quadratic to be solved. This is not the most satisfactory state of affairs. We must ask ourselves if there is not a more general way of approaching this problem that can be applied to any quadratic equation and indeed there is. Consider the general quadratic equation:

$$ax^2 + bx + c = 0$$

If we divide both sides of the equation by a, noting that $0 \div a = 0$, we produce the equation:

$$x^2 + (b/a)x + (c/a) = 0$$

To find the solution to this equation we employ the process of completing the squares. Consider the perfect square:

$$(x + [b/2a])^2 = x^2 + 2(b/2a)x + (b/2a)^2$$
$$= x^2 + (b/a)x + (b^2/4a^2)$$

Subtracting $(b^2/4a^2)$ from both sides of this equation yields:

$$(x + [b/2a])^2 - (b^2/4a^2) = x^2 + (b/a)x$$

Here, the expression on the right-hand side of this equation consists of the first two terms of the general quadratic, so if we add (c/a) to both sides of this equation to complete the quadratic on the right-hand side, we find that:

$$(x + [b/2a])^2 - (b^2/4a^2) + (c/a) = x^2 + (b/a)x + (c/a)$$

Here, the expression on the right-hand side is now equivalent to the general quadratic expression. This means that the general quadratic equation can be rewritten as:

$$(x + [b/2a])^2 - (b^2/4a^2) + (c/a) = 0$$

This means that:

$$(x + [b/2a])^2 = (b^2/4a^2) - (c/a)$$
$$= (b^2/4a^2) - (4ac/a^2)$$
$$= (b^2 - 4ac)/4a^2$$

Taking square roots of both sides of this equation gives:

$$x + [b/2a] = \pm\sqrt{[(b^2 - 4ac)/4a^2]}$$
$$= \pm\sqrt{(b^2 - 4ac)}/2a$$

Finally, subtracting $b/2a$ from both sides:

$$x = -(b/2a) \pm \sqrt{(b^2 - 4ac)}/2a$$
$$= (-b \pm \sqrt{(b^2 - 4ac)})/2a$$

In conclusion, the general quadratic equation $ax^2 + bx + c = 0$ has the two solutions:

$$x = (-b + \sqrt{(b^2 - 4ac)})/2a$$

and

$$x = (-b - \sqrt{(b^2 - 4ac)})/2a$$

Thus we have found a single method of solving any quadratic equation. For example, the equation:

$$3x^2 + 2x - 5 = 0$$

is equivalent to the general quadratic equation:

$$ax^2 + bx + c = 0$$

where $a = 3$, $b = 2$ and $c = -5$. Accordingly, the solution is given by substituting into the general form of the solution:

$$x = (-b \pm \sqrt{[(b^2 - 4ac)]})/2a$$

to give:

$$
\begin{aligned}
x &= (-2 \pm \sqrt{[2^2 - 4 \times 3 \times (-5)]})/(2 \times 3) \\
&= (-2 \pm \sqrt{[4 + 60]})/6 \\
&= (-2 \pm 8)/6 \\
&= (-2 + 8)/6 \text{ or } (-2 - 8)/6 \\
&= 1 \text{ or } -5/3
\end{aligned}
$$

The solutions $x = 1$ and $x = -5/3$ are the two solutions to the quadratic equation.

Worked Examples

8.9 Solve each of the following quadratic equations:
(a) $x^2 - x - 1 = 0$ (b) $3x^2 + 2x - 3 = 0$
(c) $x^2 - 4x + 2 = 0$ (d) $1 + 3x - x^2 = 0$
(e) $8 - 6x - 3x^2 = 0$

Solution:
(a) $x^2 - x - 1 = 0$

Here, in the general form of the quadratic:

$$ax^2 + bx + c = 0$$

$a = 1$, $b = -1$ and $c = -1$. Applying the formula:

$$x = (-b \pm \sqrt{(b^2 - 4ac)})/2a$$

we find that:

$$
\begin{aligned}
x &= \{-(-1) \pm \sqrt{[(-1)^2 - 4 \times 1 \times (-1)]}\}/(2 \times 1) \\
&= \{1 \pm \sqrt{5}\}/2
\end{aligned}
$$

(b) $3x^2 + 2x - 3 = 0$

Here, $a = 3$, $b = 2$ and $c = -3$. Applying the formula we find that:

$$x = \{-2 \pm \sqrt{[4 - 4 \times 3 \times (-3)]}\}/(2 \times 3)$$
$$= \{-2 \pm \sqrt{40}\}/6$$
$$= \{-1 \pm \sqrt{10}\}/3$$

(c) $x^2 - 4x + 2 = 0$

Here, $a = 1$, $b = -4$ and $c = 2$. Applying the formula we find that:

$$x = \{4 \pm \sqrt{[16 - 8]}\}/2$$
$$= \{4 \pm \sqrt{8}\}/2$$
$$= \{2 \pm \sqrt{2}\}/2$$

(d) $1 + 3x - x^2 = 0$

Here, $a = -1$, $b = 3$ and c = 1. Applying the formula we find that:

$$x = \{-3 \pm \sqrt{[9 + 4]}\}/2$$
$$= \{-3 \pm \sqrt{13}\}/2$$

(e) $8 - 6x - 3x^2 = 0$

Here, $a = -3$, $b = -6$ and $c = 8$. Applying the formula we find that:

$$x = \{6 \pm \sqrt{[36 + 96]}\}/(-6)$$
$$= \{6 \pm \sqrt{132}\}/(-6)$$
$$= -1 \pm (\sqrt{33})/3 \text{ because } \sqrt{132} = \sqrt{(4 \times 33)} = 2\sqrt{33}$$

Exercises

8.9 Solve each of the following quadratic equations:
    (a) $x^2 + x - 1 = 0$ (b) $2x^2 - 7x - 5 = 0$
    (c) $x^2 - 5x + 5 = 0$ (d) $3 + 4x - 2x^2 = 0$
    (e) $6 - 5x + 2x^2 = 0$

The discriminant
For the general quadratic equation:

$$ax^2 + bx + c = 0$$

with solution:

$$x = (-b \pm \sqrt{(b^2 - 4ac)})/2a$$

the quantity

$b^2 - 4ac$

in the solution is called the **discriminant** as it discriminates between three possible types of solution. The solution involves the square root of the discriminant and the nature of the solution depends upon whether this quantity is positive, zero or negative.

Positive discriminant
If the discriminant is a positive number:

$$b^2 - 4ac > 0$$

the quadratic equation has two distinct solutions corresponding to the positive and negative values of the square root of the discriminant. Graphically this corresponds to the case where the graph of:

$$y = ax^2 + bx + c$$

crosses the *x*-axis in two distinct points:

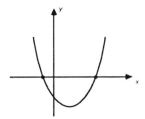

Zero discriminant
If the discriminant is zero:

$$b^2 - 4ac = 0$$

the quadratic equation has just one solution. Graphically this corresponds to the case where the graph of:

$$y = ax^2 + bx + c$$

does not cross the *x*-axis but just touches it:

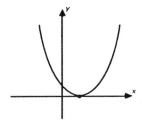

Negative discriminant
If the discriminant is a negative number:

$$b^2 - 4ac < 0$$

taking the square root of the discriminant requires us to take the square root of a negative number. As we saw in Module 3 there is no real number that is the square root of a negative number. In this case we say that the quadratic equation does not have any real solutions. Graphically, this corresponds to the case where the graph of:

$$y = ax^2 + bx + c$$

neither crosses nor touches the x-axis:

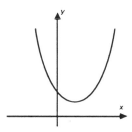

Worked Examples

8.10 Indicate whether each of the following quadratic equations has one, two or no real solutions:
(a) $x^2 + (3/2)x + 9/16 = 0$ (b) $x^2 - 0.5x - 1 = 0$
(c) $x^2 + 9x + 21 = 0$ (d) $6x^2 + x = 4$
(e) $3x^2 = 5x - 3$

Solution:
(a) $x^2 + (3/2)x + 9/16 = 0$

Here $a = 1$, $b = 3/2$, and $c = 9/16$ so that the discriminant:

$$b^2 - 4ac = 9/4 - 4 \times 9/16$$
$$= 0$$

The quadratic equation has, therefore, one real solution.

(b) $x^2 - 0.5x - 1 = 0$

Here $a = 1$, $b = -1/2$, and $c = -1$ so that the discriminant:

$$b^2 - 4ac = 1/4 + 4$$
$$> 0$$

The quadratic equation has, therefore, two real solutions.

(c) $x^2 + 9x + 21 = 0$

Here $a = 1$, $b = 9$, and $c = 21$ so that the discriminant:

$$b^2 - 4ac = 81 - 84$$
$$< 0$$

The quadratic equation has, therefore, no real solution.

(d) $6x^2 + x = 4$

Here $a = 6$, $b = 1$, and $c = -4$ so that the discriminant:

$$b^2 - 4ac = 1 + 96$$
$$> 0$$

The quadratic equation has, therefore, two real solutions.

(e) $3x^2 = 5x - 3$

Here $a = 3$, $b = -5$, and $c = 3$ so that the discriminant:

$$b^2 - 4ac = 25 - 36$$
$$< 0$$

The quadratic equation has, therefore, no real solutions.

8.11 In each of the following find the value of k that ensures that the quadratic equation has at least one real solution:
(a) $x^2 + kx + 1 = 0$ (b) $kx^2 + 2x + 3 = 0$
(c) $5x^2 - 2x + k = 0$ (d) $2k - 3kx + x^2 = 0$
(e) $kx^2 + 3kx - 1 = 0$

Solution:
(a) $x^2 + kx + 1 = 0$

Here $a = 1$, $b = k$, and $c = 1$ so that the discriminant:

$$b^2 - 4ac = k^2 - 4$$

Consequently, for the quadratic equation to have at least one real solution, it is necessary that:

$$k^2 - 4 \geq 0$$

that is:

$$k^2 \geq 4$$

which means that:

$$k \geq 2 \text{ or } k \leq -2$$

(b) $kx^2 + 2x + 3 = 0$

Here $a = k$, $b = 2$, and $c = 3$ so that the discriminant:

$$b^2 - 4ac = 4 - 12k$$

Consequently, for the quadratic equation to have at least one real solution, it is necessary that:

$$4 - 12k \geq 0$$

that is:

$$12k \leq 4$$

which means that:

$$k \leq 1/3$$

(c) $5x^2 - 2x + k = 0$

Here $a = 5$, $b = -2$, and $c = k$ so that the discriminant:

$$b^2 - 4ac = 4 - 20k$$

Consequently, for the quadratic equation to have at least one real solution, it is necessary that:

$$4 - 20k \geq 0$$

that is:

$$20k \leq 4$$

which means that:

$$k \leq 1/5$$

(d) $2k - 3kx + x^2 = 0$

Here $a = 1$, $b = -3k$, and $c = 2k$ so that the discriminant:

$$b^2 - 4ac = 9k^2 - 8k$$

Consequently, for the quadratic equation to have at least one real solution, it is necessary that:

$$9k^2 - 8k = k(9k - 8) \geq 0$$

that is:

$$(k \geq 0 \text{ and } 9k - 8 \geq 0) \; or \; (k < 0 \text{ and } 9k - 8 < 0)$$

which means that:

$$(k \geq 0 \text{ and } k \geq 8/9) \; or \; (k < 0 \text{ and } k < 8/9)$$

that is:

$$k \geq 8/9 \text{ or } k < 0$$

(e) $kx^2 + 3kx - 1 = 0$

Here $a = k$, $b = 3k$, and $c = -1$ so that the discriminant:

$$b^2 - 4ac = 9k^2 + 4k$$

Consequently, for the quadratic equation to have at least one real solution, it is necessary that:

$$9k^2 + 4k = k(9k + 4) \geq 0$$

that is:

$$(k \geq 0 \text{ and } 9k + 4 \geq 0) \; or \; (k < 0 \text{ and } 9k + 4 < 0)$$

which means that:

$$(k \geq 0 \text{ and } k \geq -4/9) \; or \; (k < 0 \text{ and } k < -4/9)$$

that is:

$$k \geq 0 \text{ or } k < -4/9$$

8.12 Show that if $x = h$ and $x = k$ are the solutions of the quadratic equation:

$$ax^2 + bx + c = 0$$

then:

$$h + k = -b/a \text{ and } hk = c/a$$

Solution:
From the formula we can say that:

$$h = (-b + \sqrt{[b^2 - 4ac]})/2a \text{ and } k = (-b - \sqrt{[b^2 - 4ac]})/2a$$

so that:

$$\begin{aligned}
h + k &= (-b + \sqrt{[b^2 - 4ac]})/2a + (-b - \sqrt{[b^2 - 4ac]})/2a \\
&= -b/2a - b/2a \\
&= -b/a
\end{aligned}$$

Also:

$$\begin{aligned}
hk &= (-b + \sqrt{[b^2 - 4ac]})/2a \times (-b - \sqrt{[b^2 - 4ac]})/2a \\
&= (-b/2a)^2 - (\sqrt{[b^2 - 4ac]}/2a)^2 \\
&= b^2/4a^2 - [b^2 - 4ac]/4a^2 \\
&= b^2/4a^2 - b^2/4a^2 + 4ac/4a^2 \\
&= c/a
\end{aligned}$$

Exercises

8.10 Indicate whether each of the following quadratic equations has one, two or no real solutions:
(a) $x^2 + (2/3)x + 1/9 = 0$ (b) $8x^2 + 0.25x + 9 = 0$
(c) $x^2 + 7x + 12 = 0$ (d) $4x^2 - x = 5$
(e) $3x^2 = 5x - 2$

8.11 In each of the following find the value of k that ensures that the quadratic equation has at least one real solution:
(a) $x^2 + kx + 2 = 0$ (b) $kx^2 - x + 4 = 0$
(c) $3x^2 + 5x - k = 0$ (d) $3k + 2kx + 3x^2 = 0$
(e) $kx^2 - 2kx + 1 = 0$

8.12 Show that if $x = h$ and $x = k$ are the solutions of the quadratic equation:

$$x^2 + x - 1 = 0$$

then:

$$hk = -(h + k)$$

Module 8   Further exercises

1   Transpose each of the following equations:
(a) $y = -9x - 11$
(b) $p = -4 - 3q^2$
(c) $a = (3b - 4)^3 - 9$
(d) $y = [(2x^{1/3}) - 3]^{1/4}$
(e) $u = [-11 - (v^2 - 5)^{-2}]^{-3}$

2   From each of the following equations obtain two alternative forms where each alternative form has a dependent variable isolated on the left-hand side:
(a) $2x - 3y = 18$
(b) $x^2 + y^2 = 2$
(c) $(a/4)^2 + (b/5)^2 = 8$
(d) $(p/7)^2 - (q/9)^2 = 4$

3   Find the range of values of the independent variable for which the dependent variable has real values in each of the following:
(a) $x^2 + y^2 = 2$
(b) $(a/4)^2 + (b/5)^2 = 8$
(c) $(p/7)^2 - (q/9)^2 = 4$

4   In each of the following, write the polynomial in ascending or descending order, give the order of the polynomial, and list the coefficients in descending power order:
(a) $-2 + 3x^2 - 4x$
(b) $x - 1 + x^2 - x^3$
(c) $8 - 3x^2$
(d) $-14 + 2x^4$
(e) $x^3 - 8$

5   Solve each of the following linear equations:
(a) $y = 4x + 3$ when $y = 9$
(b) $y = -8 + 9x$ when $y = -3$
(c) $2x + 8y = 7$ when $y = -2$
(d) $3x - 6 = 0$
(e) $5 - 4x = 0$

6   Use transposition to solve each of the following pairs of simultaneous equations:
(a) $7x - 9y = 6$
    $4x - y = -5$
(b) $13x + y = 1$
    $y = 12x - 24$

(c) $254x^2 - y^2 = 15$
    $5x + y = 6$
(d) $y^2 + y - x^2 = 1$
    $2y - 3x = 4$

7   Solve each of the following quadratic equations:
(a) $y = x^2 + 3x - 4$ when $y = 2$
(b) $y = 65 - 11x + 22x^2$ when $y = -1$
(c) $y = 2x^2 - 10x + 11$ when $y = 3$
(d) $y = 8x^2 + 6x - 9$ when $y = -10$
(e) $x^2 + x - 20 = 0$

8   Solve each of the following quadratic equations by completing the squares:
     (a) $x^2 + 5x + 1 = 0$ (b) $x^2 - 4x - 8 = 0$
     (c) $x^2 + 3x/2 - 1/4 = 0$ (d) $2x^2 + x - 1 = 0$
     (e) $1 - 4x - x^2 = 0$

9   Solve each of the following quadratic equations:
     (a) $x^2 + 3x - 1 = 0$ (b) $2x^2 + 5x - 1 = 0$
     (c) $x^2 + 6x - 6 = 0$ (d) $2 - 3x - 3x^2 = 0$
     (e) $1 + x - x^2 = 0$

10   Indicate whether each of the following quadratic equations has one, two or no
     real solutions:
     (a) $x^2 + 6x + 9 = 0$ (b) $2x^2 - 7x + 4 = 0$
     (c) $x^2 + x + 1 = 0$ (d) $2x^2 + x = 1$
     (e) $x^2 = 3x - 1$

11   In each of the following find the value of k that ensures that the quadratic
     equation has at least one real solution:
     (a) $x^2 + kx + 1 = 0$ (b) $kx^2 - 3x + 5 = 0$
     (c) $4x^2 - 7x - k = 0$ (d) $4k - kx + 5x^2 = 0$
     (e) $kx^2 - 3kx + 1 = 0$

12   Show that if $x = h$ and $x = h + 1$ are the solutions of the quadratic equation:

$$ax^2 + bx + c = 0$$

     then:

$$h^2 + 3h + 1 = (c - b)/a$$

Module 9

Division and factors

OBJECTIVES

When you have completed this module you will be able to:

- Divide two polynomials

- Use the Factor Theorem and division to factorize a polynomial

---

There are two units in this module:

Unit 1: Polynomial division
Unit 2: The Factor Theorem

Unit 1   Polynomial division

Try the following test:

1   Perform each of the following divisions:
    (a) $(x^2 + x - 2) \div (x - 1)$
    (b) $(x^3 + 6x^2 + 11x + 6) \div (x + 3)$
    (c) $(6x^3 + 11x^2 + x - 4) \div (2x^2 + x - 1)$
    (d) $(3x^2 + 6x + 9) \div (x - 5)$
    (e) $(8x^3 - 4x^2 + 2x - 1) \div (x + 2)$

Division

The process of division is one of repetitive subtraction. To perform the division:

$$(x^2 + 5x + 6) \div (x + 3)$$

we note that we are required to find out how many times $(x + 3)$ can be subtracted from $x^2 + 5x + 6$. Consider the leading term x^2 in the numerator and the leading term x in the denominator. Because:

$$x + x + \ldots + x \ (x \text{ times}) = (x) \times (x) = x^2$$

we see that x can be subtracted x times from x^2. We are not just subtracting x but $x + 3$ so we proceed as follows:

$$(x^2 + 5x - 6) - x(x + 3) = x^2 + 5x + 6 - x^2 - 3x$$
$$= 2x + 6$$

Subtracting $x - 3$ from $x^2 + 5x + 6$ x times leaves a remainder $2x + 6$. Now we subtract $x + 3$ from this remainder, noting the x can be subtracted 2 times from $2x$:

$$(2x + 6) - 2(x + 3) = 2x + 6 - 2x - 6$$
$$= 0$$

Hence:

$$x^2 + 5x + 6 - x(x + 3) + 2(x + 3) = 0$$

that is:

$$x^2 + 5x + 6 - (x + 2)(x + 3) = 0$$

that is:

$x^2 + 5x + 6 = (x + 2)(x + 3)$ so that:

$$(x^2 + 5x + 6) \div (x + 3) = (x + 2)$$

There is a tabulation procedure for performing this repetitive subtraction:

$$
\begin{array}{r}
x + 2 \\
\hline
x + 3 \,\overline{\smash{\big)}\, x^2 + 5x + 6} \\
\underline{x^2 + 3x + 0} \\
2x + 6 \\
\underline{2x + 6} \\
0
\end{array}
$$

Store the cumulative result on this line
Divide the leading term x^2 by x to give x
Multiply $x + 3$ by x and subtract from $x^2 + 5x + 6$
Divide the leading term $2x$ by x to give **2**
Multiply $x + 3$ by 2 and subtract from $2x + 6$
Remainder 0

Worked Examples

9.1 Perform each of the following divisions:

(a) $(x^2 + 5x + 4) \div (x + 1)$ (b) $(x^3 - 3x^2 + 3x - 1) \div (x - 1)$
(c) $(2x^3 - 3x^2 - 11x + 6) \div (x^2 - x - 6)$ (d) $(4x^3 + 5x + 6) \div (x - 3)$
(e) $(3x^4 + x^2 - 64) \div (x - 4)$

Solution:
(a) $(x^2 + 5x + 4) \div (x + 1)$

$$
\begin{array}{r}
x + 4 \\
\hline
x + 1 \,\overline{\smash{\big)}\, x^2 + 5x + 4} \\
\underline{x^2 + x + 0} \\
4x + 4 \\
\underline{4x + 4} \\
0
\end{array}
$$

Note that the zero must be inserted to maintain the column structure.

Therefore, $(x^2 + 5x + 4) \div (x + 1) = (x + 4)$

(b) $(x^3 - 3x^2 + 3x - 1) \div (x - 1)$

$$
\begin{array}{r}
x^2 - 2x + 1 \\
\hline
x - 1 \,\overline{\smash{\big)}\, x^3 - 3x^2 + 3x - 1} \\
\underline{x^3 - x^2 + 0 + 0} \\
-2x^2 + 3x - 1 \\
\underline{-2x^2 + 2x + 0} \\
x - 1 \\
\underline{x - 1} \\
0
\end{array}
$$

Therefore, $(x^3 - 3x^2 + 3x - 1) \div (x - 1) = (x^2 - 2x + 1)$

(c) $(2x^3 - 3x^2 - 11x + 6) \div (x^2 - x - 6)$

$$
\begin{array}{r}
2x - 1 \\
x^2 - x - 6 \enclose{longdiv}{\,2x^3 - 3x^2 - 11x + 6} \\
\underline{2x^3 - 2x^2 - 12x + 0} \\
-x^2 + \quad x + 6 \\
\underline{-x^2 + \quad x + 6} \\
0
\end{array}
$$

Therefore, $(2x^3 - 3x^2 - 11x + 6) \div (x^2 - x - 6) = (2x - 1)$

(d) $(4x^3 + 5x + 6) \div (x - 3)$

$$
\begin{array}{r}
4x^2 + 12x + 41 \\
x - 3 \enclose{longdiv}{\,4x^3 + 0x^2 + 5x + \quad 6} \\
\underline{4x^3 - 12x^2 + 0 + \quad 0} \\
12x^2 + 5x + \quad 6 \\
\underline{12x^2 - 36x + \quad 0} \\
41x + \quad 6 \\
\underline{41x - 123} \\
+129
\end{array}
$$

Therefore, $(4x^3 + 5x + 6) \div (x - 3) = (4x^2 + 12x + 41)$ remainder 129.

That is $(4x^3 + 5x + 6) \div (x - 3) = (4x^2 + 12x + 41) + 129/(x - 3)$

(e) $(3x^4 + x^2 - 64) \div (x - 4)$

$$
\begin{array}{r}
3x^3 \quad +12x^2 \quad + 49x \quad +156 \\
x - 4 \enclose{longdiv}{\,3x^4 + \quad 0x^3 \quad + \quad x^2 \quad + \quad 0x \quad - \quad 64} \\
\underline{3x^4 - \quad 12x^3 \quad + 0x^2 \quad + \quad 0x \quad + \quad 0} \\
12x^3 \quad + \quad x^2 \quad + \quad 0x \quad - \quad 64 \\
\underline{12x^3 \quad -48x^2 \quad + \quad 0x \quad + \quad 0} \\
49x^2 \quad + \quad 0x \quad - \quad 64 \\
\underline{49x^2 \quad -156x \quad + \quad 0} \\
156x \quad - \quad 64 \\
\underline{156x \quad -624} \\
560
\end{array}
$$

Therefore, $(3x^4 + x^2 - 64) \div (x - 4) = 3x^3 + 12x^2 + 49x + 156 + 560/(x - 4)$

Exercises

9.1 Perform each of the following divisions:

(a) $(x^2 + 3x + 2) \div (x + 2)$ (b) $(x^3 + 9x^2 + 27x + 81) \div (x + 3)$

(c) $(6x^3 - x^2 - 4x - 1) \div (3x^2 - 2x - 1)$ (d) $(5x^2 - 9x + 12) \div (x + 4)$

(e) $(4x^4 + 3x^2 + 5x) \div (x + 2)$

Unit 2   The Factor Theorem

Try the following test:

1   Factorize each of the following polynomials:
    (a) $x^2 + x - 20$ (b) $3x^3 + 6x^2 - 15x - 18$
    (c) $x^4 + 3x^2 - 4$ (d) $6x^3 + 11x^2 + x - 4$
    (e) $8x^4 - 14x^3 - 9x^2 + 11x - 2$

Polynomial equations

In the previous chapter we saw that the solution to the general quadratic equation:

$$ax^2 + bx + c = 0$$

is given as:

$$x = \frac{-b \pm \sqrt{b^2 - 4ac}}{2a}$$

Remarkably, this result was known to the Babylonian mathematicians of some 4000 years ago, though maybe not in quite the form that it is presented here. Since that time many people have tried to extend this idea of a general solution to higher order polynomial equations. But it was not until the 1500s that the problem of finding the solution to a restricted form of the general cubic was solved by an Italian mathematician *Nicolo of Brescia*, also known as *Tartaglia*. The details of this and other general solutions are outside the scope of this book. Instead, we shall resort to a numerical method using the Factor Theorem.

Polynomial expressions

We already know that if two linear expressions are multiplied together the result is a quadratic. For example:

$$(x + 2)(x + 3) = x(x + 3) + 2(x + 3)$$
$$= x^2 + 3x + 2x + 6$$
$$= x^2 + 5x + 6$$

which is a polynomial of degree 2 – a quadratic. Each of the linear expressions that goes up to make the quadratic is called a **factor** of the quadratic because they each divide into the quadratic:

$$(x^2 + 5x + 6)/(x + 2) = (x + 2)(x + 3)/(x + 2)$$
$$= x + 3$$

and

$$(x^2 + 5x + 6)/(x + 3) = (x + 2)(x + 3)/(x + 3)$$
$$= x + 2$$

Similarly, a product of three linear expressions results in a cubic. For example:

$$(x + 1)(x + 2)(x + 3) = (x + 1)(x^2 + 5x + 6)$$
$$= x(x^2 + 5x + 6) + 1(x^2 + 5x + 6)$$
$$= x^3 + 5x^2 + 6x + x^2 + 5x + 6$$
$$= x^3 + 6x^2 + 11x + 6$$

Again, the three linear expressions $(x + 1)$, $(x + 2)$ and $(x + 3)$ are each factors of the cubic.

When a polynomial expression is written as a product of its factors the values of the variable for which the factors are zero are called the **roots** of the polynomial. For example, the polynomial equation:

$$x^3 + 6x^2 + 11x + 6 = 0$$

can also be written as a product of its factors:

$$(x + 1)(x + 2)(x + 3) = 0$$

Consequently, the values of x which satisfy this equation are:

$$x = -1, x = -2 \text{ and } x = -3$$

as these are the values at which the factors are respectively zero. These solutions of the polynomial equation are the roots of the polynomial expression. This fact is the central theme of the **Factor Theorem**.

The Factor Theorem
If a polynomial expression in the single variable x has the value zero when x has the value a then $(x - a)$ is a factor of the polynomial. We shall not prove this statement but will proceed by example. For example, if we evaluate the polynomial:

$$x^3 + 4x^2 - x - 4$$

for $x = 1$ we see that:

$$1 + 4 - 1 - 4 = 0$$

Therefore $(x - 1)$ is a factor of the polynomial $x^3 + 4x^2 - x - 4$.

Again, if:

$$x = -1$$

the polynomial evaluates to:

$$(-1)^3 + 4(-1)^2 - (-1) - 4 = -1 + 4 + 1 - 4$$
$$= 0$$

Consequently $(x - (-1)) = (x + 1)$ is a factor of the polynomial $x^3 + 4x^2 - x - 4$.

Finally, if $x = -4$, the polynomial evaluates to:

$$(-4)^3 + 4(-4)^2 - (-4) - 4 = -64 + 64 + 4 - 4$$
$$= 0$$

so that $(x - (-4)) = (x + 4)$ is also a factor of the polynomial $x^3 + 4x^2 - x - 4$.

We have found three factors and a cubic has no more, therefore:

$$x^3 + 4x^2 - x - 4 = (x - 1)(x + 1)(x + 4)$$

which can easily be verified by multiplying out the three factors.

Worked Examples

9.2 Factorize each of the following polynomials:
    (a) $x^2 + 2x - 8$ (b) $3x^3 + 12x^2 - 33x - 90$
    (c) $x^4 - 5x^2 + 4$ (d) $6x^3 - 7x^2 - 7x + 6$
    (e) $6x^4 + 31x^3 + 57x^2 + 44x + 12$

Solution:
(a) $x^2 + 2x - 8$

When $x = 2$ the polynomial evaluates to:

$$4 + 4 - 8 = 0$$

therefore $x - 2$ is a factor. When $x = -4$ the polynomial evaluates to:

$$16 - 8 - 8 = 0$$

therefore $x - (-4) = x + 4$ is a factor. The resultant factorization is, therefore:

$$x^2 + 2x - 8 = (x - 2)(x + 4)$$

(b) $3x^3 + 12x^2 - 33x - 90$

When $x = -2$ the polynomial evaluates to:

$$-24 + 48 + 66 - 90 = 0$$

therefore $x + 2$ is a factor. When $x = 3$ the polynomial evaluates to:

$$81 + 108 - 99 - 90 = 0$$

therefore $x - 3$ is a factor. When $x = -5$ the polynomial evaluates to:

$$-375 + 300 + 165 - 90 = 0$$

therefore $x + 5$ is a factor. Noting that the coefficient of each term in the cubic has a common factor of 3 we can write the resultant factorization as:

$$3x^3 + 12x^2 - 33x - 90 = 3(x^3 + 4x^2 - 11x - 30)$$
$$= 3(x + 2)(x - 3)(x + 5)$$

(c) $x^4 - 5x^2 + 4$

When $x = \pm 1$ the polynomial evaluates to:

$$1 - 5 + 4 = 0$$

therefore $x + 1$ and $x - 1$ are factors. When $x = \pm 2$ the polynomial evaluates to:

$$16 - 20 + 4 = 0$$

therefore $x + 2$ and $x - 2$ are factors. Therefore:

$$x^4 - 5x^2 + 4 = (x + 1)(x - 1)(x + 2)(x - 2)$$

(d) $6x^3 - 7x^2 - 7x + 6$

When $x = -1$ the polynomial evaluates to:

$$-6 - 7 + 7 + 6 = 0$$

therefore $x + 1$ is a factor. There are no other integer values of x for which the cubic evaluates to zero. We now divide the cubic by the factor that we have found:

$$(6x^3 - 7x^2 - 7x + 6) \div (x + 1) = 6x^2 - 13x + 6$$
$$= (2x - 3)(3x - 2)$$
Therefore:

$$6x^3 - 7x^2 - 7x + 6 = (x + 1)(2x - 3)(3x - 2)$$

(e) $6x^4 + 31x^3 + 57x^2 + 44x + 12$

When $x = -1$ the polynomial evaluates to:

$$6 - 31 + 57 - 44 + 12 = 0$$

therefore $x + 1$ is a factor. When $x = -2$ the polynomial evaluates to:

$$96 - 248 + 228 - 88 + 12 = 0$$

therefore $x + 2$ is a factor. There are no other integer values of x for which the fourth order polynomial evaluates to zero. We now divide the polynomial by the product of the factors that we have found:

$$(6x^4 + 31x^3 + 57x^2 + 44x + 12) \div (x + 1)(x + 2)$$
$$= (6x^4 + 31x^3 + 57x^2 + 44x + 12) \div (x^2 + 3x + 2)$$
$$= 6x^2 + 13x + 6$$
$$= (2x + 3)(3x + 2)$$

the factorization is then:

$$6x^4 + 31x^3 + 57x^2 + 44x + 12 = (x + 1)(x + 2)(2x + 3)(3x + 2)$$

Exercises

9.2 Factorize each of the following polynomials:

(a) $x^2 - 3x - 18$

(b) $2x^3 - 6x^2 - 12x + 16$

(c) $x^4 - 4x^3 + 6x^2 - 4x + 1$

(d) $8x^3 + 34x^2 + 31x - 10$

(e) $12x^4 - 55x^3 + 52x^2 + 27x - 36$

Module 9 Further exercises

1   Perform each of the following divisions:
    (a) $(x^2 + 7x + 12) \div (x + 4)$ (b) $(x^3 + 6x^2 + 5x - 12) \div (x - 1)$
    (c) $(x^3 - 1) \div (x^2 + x + 1)$ (d) $(9x^2 - 7x + 5) \div (x - 1)$
    (e) $(2x^4 + 7x^2 - 9x) \div (2x - 1)$

2   Factorize each of the following polynomials:
    (a) $x^2 - 4x - 5$ (b) $3x^3 + 9x^2 - 99x - 105$
    (c) $x^4 - 8x^3 + 24x^2 - 32x + 16$ (d) $4x^3 - 12x^2 + 12x - 4$
    (e) $2x^4 + 20x^3 + 70x^2 + 100x + 48$

Part Three

Trigonometry

Mathematics deals with systematic developments of abstract ideas; the qualitative and quantitative description of pattern and the commonality of disparate ideas. It is intrinsically a mental activity. For reasons curious and largely unknown much of the mathematics that has evolved can be used to model the world as we perceive it. We have already seen how, in Part One, arithmetic has evolved from a desire to count physical objects to a complete abstract structure and how, in Part Two, this same mathematics could be used to model the world as we describe it in word problems. Mathematics, taken further, provides us with a language that enables us to describe and predict the behaviour of a multiplicity of events ranging from the innermost workings of the atom through to the outermost organization of the cosmos.

We live in a restricted environment embedded within a seemingly endless and infinite universe. The evidence for this is all too clear and humankind has been aware of it since recorded time began. To sit and gaze upwards at the vast array of stars on a clear, cloudless night must be one of life's most humbling experiences. Indeed, the one picture of recent times that has, perhaps, done more to put the consciousness of our finite environment into perspective is the photograph taken from the surface of the moon of our solitary earth against the deep, black backdrop of empty space. No wonder the ancients looked to the heavens in their attempts to find causes and reasons for existence; no wonder their attempts to predict and control their lives took them from their own confined environment to space and the objects contained therein.

Everything changes but some changes are cyclical and cyclical changes can be predicted. Spring, summer, autumn and winter; the perennial cycle whose repetition the ancients needed to predict so they could plan ahead; to plant and harvest crops. From both necessity and a spiritual desire to link cause and effect with the external universe, geometry came into being – the study of shape, form and comparative sizes of objects in space. The ancient Egyptians must have had a keen geometric sense to have enabled them to build their pyramids and it is known that the Babylonians had a sense of geometry expressed in algebraic terms. Geometry, as a study in its own right, came to life in Greece during the second century BC when demonstrative and deductive geometry was developed by a host of Greek mathematicians including such familiar names as *Pythagoras*, *Plato* and *Archimedes*. Trigonometry, that aspect of demonstrative geometry dealing with the properties of triangles, was initiated in the second century BC by *Hipparchus* from his attempts to quantify angular measure. A knowledge of its content coupled with an ability to manipulate its symbolism is an essential prerequisite to further study of geometric concepts.

Module 10

Angles and triangles

OBJECTIVES

When you have completed this module you will be able to:

- ■ Convert angular measure from degrees to radians and from radians to degrees

- ■ Identify different types of angle and different types of triangle

- ■ Apply the properties of right-angled triangles.

There are three units in this module:

Unit 1   Angular measure

Try the following test:

1   Convert the measure of each of the following angles into radians:
   (a) 15°
   (b) 240°
   (c) 145° 45' 45"
   (d) 720°

2   Convert the measure of each of the following angles into degrees:
   (a) $\pi/12$ r
   (b) $3\pi/5$ r
   (c) 1.018 r
   (d) 4.3625 r

3   Describe the types of angle in each of the following::

4   In each of the following indicate which angles are equal, which are complementary and which are supplementary:

5   Give an alternative measure to each of the following angles:
   (a) 45°
   (b) $\pi/2$ r
   (c) −120°
   (d) −5π/4 r

Angles

An angle is a measure of rotation. If one end point of a line segment is fixed and the line is rotated about that point the free end of the line segment traces out a circular path:

In one complete rotation the line segment returns to its original position and is said to have **rotated through a full angle.**

A full angle is subdivided into 360 **degrees** (symbolized by 360°, where the symbol ° stands for degree). The reason why 360 was selected as the number of degrees in a full rotation is not clear though there are a number of indications as to why it was chosen.

It has been suggested that the Babylonians used a unit of distance roughly equal to seven of our present miles. This distance unit was also used as a measure of time; a single time unit being the time taken to travel one Babylonian distance unit. As it took one complete day to travel 12 of these distance units and one complete day is equivalent to one complete revolution of the pattern of the stars in the sky it appears natural to divide a complete revolution into 12 equal parts, each part corresponding to one Babylonian time unit. These 12 parts were then further subdivided into 30 conveniently smaller parts resulting in the circumference of a complete circle being subdivided into $12 \times 30 = 360$ equal parts.

This division of the circumference of a circle into 360 equal parts was not at the time taken as a measure of angle; the concept of angular measure did not arise until *Hipparchus* in the second century BC.

The Greek mathematicians studied relationships between circles and line segments in their desire to predict astronomical events. One relationship in particular prompted *Hipparchus* to construct a table of the ratios of arc length to chord length:

arc AB
chord AB

both of which are created when a straight line intersects a circle:

The use of 360 divisions of a complete revolution – a full angle – arose as a result of *Hipparchus'* work which has since earned him the title of the father of trigonometry. Each of the 360 parts was then further subdivided using the Babylonian sexagesimal fractions into 60 parts and so on. Hence we obtain the division of a degree into 60 minutes (60' where the ' denotes the unit of minute) where each minute is further subdivided into 60 seconds (60" where the " denotes the unit of second).

An alternative measure of rotation is the **radian**. When the line segment has rotated to the point where the arc formed by the free end is equal to the length of the line segment the line is said to have rotated through an angle of 1 radian.

In a complete rotation the free end point of the line segment traces out the circumference of a circle whose length if $2\pi r$ where r is the length of the line segment. – the radius of the circle. Consequently there are 2π radians in one complete revolution which gives us a means whereby we can compare the two different measures of angle:

2π radians = 360 degrees

This means that:

1 radian = $(360/2\pi)$ degrees   *and*   1 degree = $(2\pi/360)$ radians
          = $(180/\pi)$ degrees   = $(\pi/180)$ radians

A third measure of angle is the **new degree** or **grade** which is defined as one four hundredth of a full angle. Again, each new degree is further subdivided, this time into 100 **new minutes**, each of which consists of 100 **new seconds**. This is not a common measure of angle, being used mainly in the military, though it is present as an option on most hand calculators.

Worked Examples

10.1 Convert the measure of each of the following angles into radians:
    (a) 45°   (b) 120°
    (c) 100° 15' 45"   (d) 450°

Solution:
(a) $45° = (45)(2\pi/360)$ r
      $= 2\pi/8$ r
      $= \pi/4$ r

(b) $120° = (120)(2\pi/360)$ r
       $= 2\pi/3$ r

(c) $100° 15' 45" = (100 + 15/60 + 45/3600)°$
                 $= (100 + 0.25 + 0.0125)°$
                 $= 100.2625°$
                 $= (100.2625)(2\pi/360)$ r
                 $= 1.7499$ r   to 4 decimal places using the
                                    approximation $\pi = 3.1415927$

(d) $450° = (450)(2\pi/360)$ r
       $= 5\pi/2$ r   **Notice**: A rotation can be greater than a full angle.

10.2 Convert the measure of each of the following angles into degrees:
    (a) $\pi/15$ r   (b) $3\pi/4$ r
    (c) 1.55 r   (d) 5.776 r

Solution:

(a) $\pi/15$ r $= (\pi/15)(360/2\pi)°$
    $= 12°$

(b) $3\pi/4$ r $= (3\pi/4)(360/2\pi)°$
    $= 135°$

(c) 1.55 r $= (1.55)(360/2\pi)°$
    $= 88.808458...°$
    $= 88°\ 48'\ 30''$ to the nearest second (") using
    $\pi = 3.1415927$

(d) 5.776 r $= (5.776)(360/2\pi)°$
    $= 330.94042...°$
    $= 330°\ 56'\ 26''$ to the nearest second (") using
    $\pi = 3.1415927$

Exercises

10.1 Convert the measure of each of the following angles into radians:
   (a) $30°$ (b) $75°$
   (c) $85°\ 25'\ 15''$ (d) $540°$

10.2 Convert the measure of each of the following angles into degrees:
   (a) $\pi/6$ r   (b) $5\pi/8$ r
   (c) 0.876 r   (d) 3.335 r

Types of angle
Certain special angles have been given specific names.

Straight angle
An angle of $180° = \pi$ radians is called a **straight angle**.

Right angle
An angle of $90° = \pi/2$ radians is called a **right angle**.

Acute angle
An angle that is less than $90° = \pi/2$ radians is called an **acute angle**.

Obtuse angle
An angle that is greater than $90° = \pi/2$ radians but less than a straight angle is called an **obtuse angle**.

Reflex angle
An angle that is greater than $180° = \pi$ radians but less than a full angle is called a **reflex angle**.

Complementary angles
If two angles add up to 90° they are said to be **complementary angles** and either one is said to be complementary to the other.

Supplementary angles
If two angles add up to 180° they are said to be **supplementary angles** and either one is said to be supplementary to the other.

Equal and opposite angles
If two lines cross they form two pairs of **opposite** angles at their point of intersection:

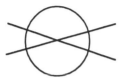

Adjacent angles are supplementary angles and opposite angles are equal.

Parallel lines
If two parallel lines are crossed by a third line, eight angles are formed, four on each side of the third line.

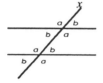

Corresponding angles
On each side of the third line the four angles consist of two pairs of **corresponding** angles and corresponding angles are equal.

Supplementary angles
On each side of the third line the four angles consist of one pair of **exterior** angles and one pair of **interior** angles. Interior angles and exterior angles are supplementary.

Alternate angles
On opposite sides of the third line are two pairs of **alternate** exterior angles and two pairs of **alternate** interior angles. Alternate angles are equal. Alternate interior angles are often termed **z-angles**:

Worked Examples

10.3 Describe the types of angle in each of the following:

Solution:
Right-angle, reflex angle, reflex angle.

10.4 In each of the following indicate which angles are equal, which are complementary and which are supplementary:

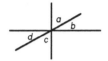

Solution:
$a = c$, $d = b$, a and b are complementary as are d and c

Exercises

10.3 Describe the types of angle in each of the following::

10.4 In each of the following indicate which angles are equal, which are complementary and which are supplementary:

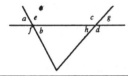

Positive and negative angles

We have stated that an angle is a measure of rotation but what we have so far failed to mention is that there are two distinct types of rotation; clockwise and anticlockwise. Not only do we need to measure the amount of rotation but we also need to indicate the type of rotation. We do this by using **positive** and **negative** angles. By convention we agree to define a positive angle as a measure of an anticlockwise rotation and to define a negative angle as a measure of a clockwise rotation. For example, in the following Figure:

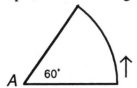

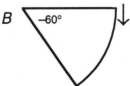

we distinguish the two angles *A* and *B* as being of value:

+60° and –60° respectively.

Notice: The use of positive and negative angles does produce a duplication of notation because a clockwise rotation of –30° could also be measured as an anticlockwise rotation of 330°. Indeed, if *x* is any measured angle then *x* – 360 is an alternative negative measure if *x* is positive and *x* + 360 is an alternative positive measure if *x* is negative.

Worked Examples

10.5 Give an alternative measure to each of the following angles:

(a) 60° (b) $2\pi/3$ r
(c) –240° (d) $-\pi/4$ r

Solution:

(a) A positive angle of 60° (anticlockwise rotation) can be alternatively measured as:

$$(60 - 360)° = -300°$$ a negative angle of –300° (clockwise rotation).

(b) A positive angle of $2\pi/3$ r (anticlockwise rotation) can be alternatively measured as:

$$(2\pi/3 - 2\pi) = -4\pi/3$$ r a negative angle of $-4\pi/3$ r (clockwise rotation).

(c) A negative angle of –240° (clockwise rotation) can be alternatively measured as:

$$(-240 + 360)° = -120°$$ a positive angle of 120° (anticlockwise rotation).

(d) A negative angle of $-\pi/4$ r (clockwise rotation) can be alternatively measured as:

$$(-\pi/4 + 2\pi) \, r = 7\pi/4 \, r$$

a positive angle of $7\pi/4$ r (anticlockwise rotation).

Exercises

10.5 Give an alternative measure to each of the following angles:

(a) 72° (b) $7\pi/5$ r
(c) –270° (d) $-\pi/6$ r

Unit 2   Triangles

Triangles

A **triangle** is a three sided, closed, **rectilinear** figure and is the simplest closed shape that can be constructed with straight line segments – a rectilinear figure is one composed of straight lines. Types of triangle can be distinguished by their sides:

Scalene triangle

If the three sides of a triangle are all different lengths the triangle is called a **scalene** triangle. In a scalene triangle all the angles are also different.

Isosceles triangle

If two of the sides of a triangle are the same length the triangle is called an **isosceles** triangle. In an isosceles triangle two of the angles are the same.

Equilateral triangle
If the three sides of a triangle are all the same length the triangle is called an **equilateral** triangle. In an equilateral triangle all three angles are the same.

Types of triangle can also be distinguished by their angles:

Acute angled triangle
If all the angles of a triangle are less than 90° ($\pi/2$ radians) the triangle is called an **acute** angled triangle

Obtuse angled triangle
If one of the angles of a triangle is greater than 90° ($\pi/2$ radians) the triangle is called an **obtuse** angled triangle

Right-angled triangle
If one of the angles of a triangle is equal to 90° ($\pi/2$ radians) the triangle is called a **right-angled** triangle

---

Worked Examples

10.6 Describe the type of each of the following triangles:

Solution:
Equilateral, right-angled, scalene

Exercises

10.6 Describe the type of each of the following triangles:

  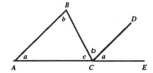

Sum of angles

The angles of a triangle add up to 180° (π radians). The proof of this statement is quite straightforward. In the following Figure:

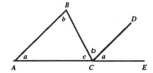

the line *CD* is drawn parallel to side *AB* of the triangle *ABC*.

Angle *BAC* is equal to angle *DCE* because they are corresponding angles
Angle *ABC* is equal to angle *BCD* because they are z-angles

The remaining angle in the triangle, angle *ACB*, then adds to the other two to form a straight angle.

$$
\begin{aligned}
\text{sum of angles in the triangle} &= ABC + BAC + ACB \\
&= BCD + DCE + ACB \\
&= 180° \text{ (straight angle)}
\end{aligned}
$$

Area of a triangle

Given the triangle *ABC*, if we construct the **altitude** *BD* which is at right angles to the side *AC* the area of the triangle is then:

$$\frac{AC \times BD}{2}$$

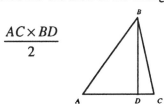

One half base times height. The proof is quite straightforward:

The area of the rectangle *AEFC* is defined as **base length** × **height**, namely:

area of $AEFC = AC \times AE$

From the Figure we can see that this area is the sum of four other areas:

area of $AEFC$ = area AEB + area ABD + area BDC + area BFC
                = 2(area ABD) + 2(area BDC)
                = 2(area ABC)

Therefore:

area $ABC = (1/2)$(area $AEFC$)
          = $(1/2)(AC \times BD)$

Worked Examples

10.7 Find the area of the following triangle:

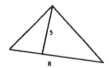

Solution:
Area = $(8 \times 5)/2 = 20$

Exercises

10.7 Find the area of the following triangle:

Unit 3   The right-angled triangle

Try the following test:

1   If ABC is a triangle right-angled at A and $AB = 15$ cm, $BC = 25$ cm calculate the length of AC.

2   Calculate the area of triangle ABC in which angle $A = 90°$, $BC = 4$ m and $AC = 2$ m.

3   Prove that a triangle with sides 14, 7, 7√3 is a right-angled triangle.

4   Find the vertical height of an equilateral triangle with side length 12 cm.

Pythagoras
Pythagoras' theorem states that:

In a right-angled triangle the square of the hypotenuse equals the sum of the squares on the other two sides.

$$a^2 + b^2 = c^2$$

An ingenious proof of this theorem was found by Abram Garfield, twentieth President of the United State of America. In the following Figure:

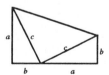

two identical right-angled triangles and a straight line are arranged to form a trapezium. The area of the trapezium is given as the average of the lengths of the two parallel sides multiplied by the perpendicular distance between them:

$$(1/2)(a + b)(a + b) = (1/2)(a^2 + 2ab + b^2)$$
$$= (1/2)(a^2 + b^2) + ab$$

This is equal to the sum of the areas of the two triangles and the half square of side c, that is:

$$(1/2)(a^2 + 2ab + b^2) = (2)(1/2)(a(b) + (1/2)c^2$$
$$= ab + (1/2)c^2$$

That is:

$$(1/2)(a^2 + b^2) = (1/2)c^2$$

so that:

$$a^2 + b^2 = c^2$$

The 3:4:5 triangle
The 3:4:5 triangle is a right-angled triangle because, by Pythagoras:

$$3^2 + 4^2 = 9 + 16$$
$$= 25$$
$$= 5^2$$

Consequently, any triangle whose sides form the ratios:

3/4, 3/5 and 4/5

is a similar triangle and is, therefore a right angled triangle. For example, the triangle with sides 6, 8 and 10 is right-angled because the sides are in the ratios:

6/8 = 3/4, 6/10 = 3/5 and 8/10 = 4/5

Worked Examples

10.8 If *ABC* is a triangle right-angled at *A* and *AB* = 12 cm, *BC* = 13 cm calculate the length of *AC*.

Solution:
By Pythagoras:

$$AB^2 + AC^2 = BC^2$$

Therefore:

$$12^2 + AC^2 = 13^2$$

That is:

$$144 + AC^2 = 169$$

therefore:

$$AC^2 = 25 \text{ and hence, } AC = 5$$

10.9 Calculate the area of triangle *ABC* in which angle $A = 90°$, $BC = 7$ m and $AC = 3$ m.

Solution:
By Pythagoras:

$$AB^2 + AC^2 = BC^2$$

Therefore:

$$AB^2 + 3^2 = 7^2$$

That is:

$$AB = \sqrt{40} = 2\sqrt{10}$$

The area of the triangle is, therefore:

$$\frac{AB \times AC}{2} = 3\sqrt{10}$$

Exercises

10.8 If *ABC* is a triangle right-angled at *A* and $AB = 13$ cm, $BC = 13\sqrt{2}$ cm calculate the length of *AC*.

10.9 Calculate the area of triangle *ABC* in which angle $A = 90°$, $BC = 12$ m and $AC = 4$ m.

Two specific triangles
Because the sides of a right-angled triangle are related by the squares of their lengths the side lengths are, in many cases, going to be represented by an irrational number. Two specific cases are worthy of mention:

The 1:1:√2 triangle
In a right-angled isosceles triangle the two other angles are both equal to $45° = \pi/4$ r:

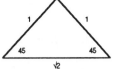 $1^2 + 1^2 = 2 = (\sqrt{2})^2$

If the two equal sides are taken to have the length 1 then the hypotenuse is of length $\sqrt{2}$.

The 1:√3:2 triangle

An equilateral triangle has three equal sides and three equal angles, each of $\pi/3\,\mathrm{r} = 60°$. If a straight line is drawn from any vertex to the opposite side to bisect the side it forms two half equilateral triangles. If the equilateral triangle is taken to have side length 2, the bisected side which forms one side of the half equilateral triangle has length 1 and the third side of the half equilateral triangle has length √3.

Worked Examples

10.10 Prove that a triangle with sides 5, 5, 5√2 is a right-angled triangle.

Solution:
$5^2 + 5^2 = 2 \times 5^2 = (5\sqrt{2})^2$ Therefore, by Pythagoras, the triangle is right-angled. Alternatively, the sides are in the ratios:

$5/5 = 1, 5/5\sqrt{2} = 1/\sqrt{2}$ and $5/5\sqrt{2} = 1/\sqrt{2}$ respectively.

The triangle is similar to the right-angled 1:1:√2 triangle

10.11 Find the vertical height of an equilateral triangle with side length 8 cm.

Solution:
The equilateral triangle with side length 8 cm is similar to the equilateral triangle with side length 2 cm and the ratio of corresponding sides is:

$8/2 = 4$

Consequently, since the vertical height of an equilateral triangle with side length 2 cm is √3 cm the vertical height of an equilateral triangle with side length 8 cm is 4√3 cm.

Exercises

10.10 Prove that a triangle with sides 2, 4, 2√3 is a right-angled triangle.

10.11 Find the side length of an equilateral triangle with vertical height 8 cm.

Module 10 Further exercises

1 Convert the measure of each of the following angles into radians:
(a) 25°   (b) 270°
(c) 203° 30' 15"   (d) 840°

2 Convert the measure of each of the following angles into degrees:
(a) π/3 r   (b) 9π/4 r
(c) 0.755 r   (d) 2.665 r

3 Describe the types of angle in each of the following::

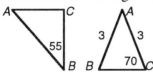

4 In each of the following indicate which angles are equal, which are
complementary and which are supplementary:

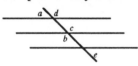

5 Give an alternative measure to each of the following angles:
(a) 90°   (b) 3π/8 r
(c) −300°   (d) −7π/6 r

6 Describe the type of each of the following triangles:

7 Find the unknown angles indicated in each of the following:

8 Find the area of the following triangle:

9   If *ABC* is a triangle right-angled at *A* and *AB* = 9 cm, *BC* = 15 cm calculate the length of *AC*.

10 Calculate the area of triangle *ABC* in which angle *A* = 90°, *BC* = 8 m and *AC* = 5 m.

11 Prove that a triangle with sides 21, 28, 35 is a right-angled triangle.

12 Find the vertical height of an equilateral triangle with side length 10 cm.

Module 11

The trigonometric ratios

OBJECTIVES

When you have completed this module you will be able to:

- ■ Use and manipulate expressions and equations involving the trigonometric ratios

- ■ Solve triangles using the sine and cosine rules

- ■ Extend the definitions of the trigonometric ratios to negative angles and to angles greater than 90°

There are three units in this module:

Unit 1: The trigonometric ratios
Unit 2: Compound angles and identities
Unit 3: The sine and cosine rules

Unit 1   The trigonometric ratios

Try the following test:

1   Find the sine, cosine and tangent of each of the acute angles in the
following triangle:

2   Use a calculator to find the sine, cosine and tangent of each of the
following angles:
(a) 35°   (b) 1.35 r
(c) 4π/9 r   (d) 89.637°

3   Find the value of the secant, cosecant and cotangent of each of the
following angles:
(a) 67°   (b) 0.75 r
(c) 3π/7 r   (d) 32.154°

Similarity

The ability to depict the exterior world in picture form is an attribute of nature that
has been familiar to us since the time when humans first painted images of their
experiences on the walls of their caves. So familiar is it that when we look at the
primitive cave drawings of our hominid ancestors we immediately recognize the
shapes and forms of animals that still exist to this day. What is it about such pictures
that even though the dimensions have been scaled down to fit a cave wall we can still
recognize the shapes? Indeed, what is it about any picture that enables us to recognize
what it is purporting to portray?

The secret of drawing a recognizable shape whilst scaling the dimensions up or
down lies in the preservation of the angular structure of the figure being drawn. For
example, if you look at a map of your locality all the distances are naturally scaled
down but the directions – the angles – are maintained. The map is not the same as the
locality it represents in the sense of being identical, because the actual distances in
the locality are obviously much larger than the actual distances on the map. However,
by preserving directions, that is angles, it is **similar** and this similarity is what makes
it recognizable as a map of your locality.

The need to formalize this concept of similarity is evident whenever we wish to make
a drawing. Be it something to hang on your wall in a frame or a scaled plan of a
pyramid in which to inter the body of a Pharaoh it will require a formal use of
similarity for its portrayal to be recognizable.

Similar triangles

Two triangles are said to be similar if their corresponding angles are equal. The
immediate effect of this is that pairs of corresponding sides of two similar triangles

are in constant ratio. For example, triangles ABC and $A'B'C$ of the following Figure are similar so that:

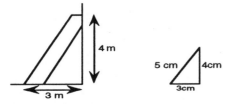

$$\frac{AB}{A'B} = \frac{AC}{A'C} = \frac{BC}{B'C}$$

From these ratios we can deduce the equality of ratios in one triangle to the corresponding ratios in the second triangle.

$$\frac{AB}{AC} = \frac{A'B}{A'C}$$

$$\frac{AB}{BC} = \frac{A'B}{B'C}$$

$$\frac{AC}{BC} = \frac{A'C}{B'C}$$

These latter three equalities are basic to trigonometry. For example, a wall has to be supported by a wooden prop as shown:

It is known that the height of the vertical strut is 4 m and the length of the horizontal strut is 3 m but it is not known how long the leaning strut will be. If a diagram is drawn of this prop scaled down so that 1 cm on the drawing represents 1 m in reality then it is found that measuring the leaning line in the diagram gives 5 cm. From the drawing we can derive three ratios of side lengths:

3/4, 3/5 and 4/5 respectively

Because the triangle in the drawing and the triangle of the prop are similar we can say that these ratios are preserved – the lengths of the sides of the prop are in the ratios:

3/4, 3/5 and 4/5 correspondingly

Hence the length of the leaning part of the wooden strut is 5 m. Here we can see that a knowledge of the ratios of the sides of a triangle will permit deductions to be made about all triangles that are similar to the original. These ratios are called the **trigonometric** ratios and are defined using a right-angled triangle.

The trigonometric ratios

The right-angled triangle in the following Figure possesses a right angle and an angle labelled x :

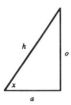

The length of the side opposite angle x is labelled o, the length of the side adjacent to angle x is labelled a and the length of the third side, called the hypotenuse, is labelled h. The three trigonometric ratios are then defined as:

■   **sine**
    The **sine** of the angle x is defined as the ratio o/h, written as:

$$\sin(x) = o/h$$

■   **cosine**
    The **cosine** of the angle x is defined as the ratio a/h, written as:

$$\cos(x) = a/h$$

■   **tangent**
    The **tangent** of the angle x is defined as the ratio of the sine and the cosine and is written as:

$$\tan(x) = \sin(x)/\cos(x)$$
$$= (o/h)/(a/h)$$
$$= o/a$$

Notation
Usually, the brackets are omitted and instead of writing:

$$\sin(x), \cos(x) \text{ and } \tan(x)$$

we simply write:

$$\sin x, \cos x \text{ and } \tan x$$

We shall reserve the right to use either notation at will. Also, when we raise a trigonometric ratio to a power, for example:

$$[\sin x]^n$$

we use the notation:

$$\sin^n x$$

Worked Examples

11.1 Find the sine, cosine and tangent of each of the acute angles in the
following triangle:

Solution:
$A = C = 45°$: $\sin A = \sin C = \cos A = \cos C = 1/\sqrt{2}$, $\tan A = \tan C = 1$

Exercises

11.1 Find the sine, cosine and tangent of each of the acute angles in the
following triangle:

Using a calculator
On a hand calculator with mathematical functions you will find the three trigonometric ratios amongst the function keys. These are used to find the trigonometric ratios of the angle displayed on the calculator screen. However, because angles can be measured in either degrees or radians the calculator must first be put into the appropriate **mode**, displayed on the screen as either DEG or RAD – the calculator manual will tell you how to change from one mode to another.

Put the calculator in DEG mode, enter the number 2 and press the **sin** function key to display the number:

0.0348994...

This is the value of the sine of 2 degrees – sin 2°. Clear the display, change the mode to RAD, enter the number 2 and press the **sin** function key to display the number:

0.9092974...

This is the value of the sine of 2 radians – sin 2 r.

Worked Examples

11.2 Use a calculator to find the sine, cosine and tangent of each of the following angles:

(a) 23°   (b) 0.24 r
(c) 6π/17 r   (d) 74.012°

Solution:

(a) sin 23° = 0.3907..., cos 23° = 0.9205..., tan 23° = 0.4244...

(b) sin 0.24 r = 0.2377..., cos 0.24 r = 0.9713..., tan 0.24 r = 0.2447...

(c) sin 6π/17 r = 0.8951..., cos 6π/17 r = 0.4457..., tan 6π/17 r = 2.0082...

(d) sin 74.012° = 0.9613..., cos 74.012° = 0.2754..., tan 74.012° = 3.4901...

Exercises

11.2 Use a calculator to find the sine, cosine and tangent of each of the following angles:

(a) 46°   (b) 1.13 r
(c) 3π/8 r   (d) 63.175°

The reciprocal ratios
Each trigonometric ratio has its defined reciprocal:

■ secant   The **secant** of the angle x is defined as the reciprocal of the cosine ratio and is written as:

$$\sec x = 1/\cos x$$

■ cosecant   The **cosecant** of the angle x is defined as the reciprocal of the sine ratio and is written as:

$$\operatorname{cosec} x = 1/\sin x$$

■ cotangent   The **cotangent** of the angle x is defined as the reciprocal of the tangent ratio and is written as:

$$\cot x = 1/\tan x$$

Complementary angles
In a right-angled triangle the two acute angles add up to 90° which means that each angle is complementary to the other. For this reason the cosine of an angle is equal to the sine of its complement:

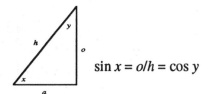

$$\sin x = o/h = \cos y$$

Similarly, the cosecant of an angle is the secant of the angle's complement and the cotangent of an angle is the tangent of the angle's complement. The names tangent and secant are derived from their immediate geometric counterparts but the name sine has a most curious history. In the following Figure a circle of radius 1 has been drawn:

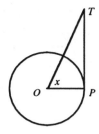

The line *PT* is tangential to the circle at *P* and, because *OP* = 1, its length is equal to the ratio:

$$\frac{PT}{OP}$$

the trigonometric ratio tan *x*. The line *OT* is called a **secant** – a straight line that intersects a curve – and its length is equal to the ratio:

$$\frac{OT}{OP}$$

the reciprocal trigonometric ratio sec *x*. The line *ST* is a half-chord and its length is equal to the ratio:

$$\frac{ST}{OP}$$

the trigonometric ratio sin *x*. The lengths of half-chords *ST* for different angles *x* were the subject of *Hipparchus'* tabulation when trigonometry was being developed. When the work of the Greek mathematicians was being translated by Arab scholars the half-chord took on its Arabian equivalent **jya-ardha** which was abbreviated to **jya**. Over time, presumably by a process of Chinese whispers coupled with the omission of vowels in the written word this word mutated first to a nonsensical word **jiba** – written without the vowels as **jb** – and then to **jaib** which means **bay**. In the twelfth century, when European scholars were translating Arabic translations of Greek works into Latin **jaib** was translated to the Latin for bay which is **sinus**. Hence the name of the trigonometric ratio **sin**.

Worked Examples

11.3 Find the value of the secant, cosecant and cotangent of each of the following
angles:
   (a) 18°   (b) 1.47 r
   (c) 9π/20 r   (d) 33.333°

Solution:
(a) cos 18° = 0.9510... and sec 18° = 1/(cos 18°) = 1.0514...
     sin 18° = 0.3090... and cosec 18° = 1/(sin 18°) = 3.2360...
     tan 18° = 0.3249... and cot 18° = 1/(tan 18°) = 3.0776...

(b) cos 1.47 r = 0.1006... and sec 1.47 r = 1/(cos 1.47 r) = 9.9378...
     sin 1.47 r = 0.9949... and cosec 1.47 r = 1/(sin 1.47 r) = 1.0051...
     tan 1.47 r = 9.8873... and cot 1.47 r = 1/(tan 1.47 r) = 0.1011...

(c) cos 9π/20 r = 0.1564... and sec 9π/20 r = 1/(cos 9π/20 r) = 6.3924...
     sin 9π/20 r = 0.9876... and cosec 9π/20 r = 1/(sin 9π/20 r) = 1.0124...
     tan 9π/20 r = 6.3137... and cot 9π/20 r = 1/(tan 9π/20 r) = 0.1583...

(d) cos 33.333° = 0.8354... and sec 33.333 ° = 1/(cos 33.333 °) = 1.1969...
     sin 33.333 ° = 0.5495... and cosec 33.333 ° = 1/(sin 33.333 °) = 1.8198...
     tan 33.333 ° = 0.6577... and cot 33.333 ° = 1/(tan 33.333 °) = 1.5204...

Exercises

11.3 Find the value of the secant, cosecant and cotangent of each of the following
angles:
   (a) 58°   (b) 0.015 r
   (c) 13π/24 r   (d) 4.326°

Unit 2   Compound angles and identities

Try the following test:

1. Expand each of the following:
   (a) $\sin(x + \pi/3)$ (b) $\cos(x - \pi/4)$
   (c) $\tan(x - \pi/6)$ (d) $\sin 2(x + \pi/12)$

2. Without using a calculator evaluate:
   (a) $\cos 15°$ (b) $\tan 5\pi/12$

3. Show that:
   (a) $\cos 45° + \cos 15° = \sqrt{3}\cos 15°$
   (b) $\sin 45° - \sin 15° = \sqrt{3}\sin 15°$

4. Expand $\cos 4x$ in terms of $\cos x$

Compound angles

The trigonometric ratios of sums and differences of angles form that aspect of trigonometry called the trigonometry of compound angles. Here we shall derive the sine of a sum of angles and merely quote all the other combinations. In the following Figure:

PS and QT are perpendicular to OT
RQ is parallel to OT
Angle SOQ = angle OQR = angle QPR = x
RS = QT.

From the Figure we see that:

$$\sin(x + y) = SP/OP$$
$$= (SR + RP)/OP$$
$$= (TQ + RP)/OP$$
$$= TQ/OP + RP/OP$$
$$= (TQ/OQ)(OQ/OP) + (RP/QP)(QP/OP)$$
$$= \sin x \cos y + \cos x \sin y$$

The complete collection of compound angle formulae is:

The sine of a sum or difference

$$\sin(x + y) = \sin x \cos y + \sin y \cos x$$
$$\sin(x - y) = \sin x \cos y - \sin y \cos x$$

In particular, if $x = y$ then:

$$\sin 2x = 2\sin x \cos x$$

The cosine of a sum or difference

$$\cos(x + y) = \cos x \cos y - \sin y \sin x$$
$$\cos(x - y) = \cos x \cos y + \sin y \sin x$$

In particular, if $x = y$ then:

$$\cos 2x = \cos x \cos x - \sin x \sin x$$

that is:

$$\cos 2x = \cos^2 x - \sin^2 x$$

The tangent of a sum or difference

$$\tan(x + y) = (\tan x + \tan y)/(1 - \tan x \tan y)$$
$$\tan(x - y) = (\tan x - \tan y)/(1 + \tan x \tan y)$$

In particular, if $x = y$ then:

$$\tan 2x = (2\tan x)/(1 - \tan^2 x)$$

The sum or difference of sines

$$\sin x + \sin y = 2\sin[(x + y)/2] \cos[(x - y)/2]$$
$$\sin x - \sin y = 2\sin[(x - y)/2] \cos[(x + y)/2]$$

The sum or difference of cosines

$$\cos x + \cos y = 2\cos[(x + y)/2] \cos[(x - y)/2]$$
$$\cos x - \cos y = -2\sin[(x - y)/2] \sin[(x + y)/2]$$

Notice: Each of these formulae is a special kind of equation; special in the sense that the equality holds true for **any** value of the variables and not just for specific values. Any equation of this type is called an **identity**. The symbol for equality in an identity is $\equiv$ so that the equation:

$$\cos x - \cos y = -2\sin[(x - y)/2] \sin[(x + y)/2]$$

should be written as:

$$\cos x - \cos y \equiv -2\sin[(x - y)/2] \sin[(x + y)/2]$$

However, it is common practice to use the = sign in place of the less familiar $\equiv$ sign.

Worked Examples

11.4 Expand each of the following:
  (a) $\cos(x + \pi/4)$ (b) $\sin(x + \pi/6)$
  (c) $\tan(x + \pi/3)$ (d) $\cos 3(x - \pi/18)$

Solution:
(a) Using the cosine of a sum of two angles formula:

$$\cos(x + \pi/4) = \cos x \cos \pi/4 - \sin x \sin \pi/4$$
$$= (1/\sqrt{2})(\cos x - \sin x) \qquad \text{since } \cos \pi/4 = \sin \pi/4 = 1/\sqrt{2}$$

(b) Using the sine of a sum of two angles formula:

$$\sin(x + \pi/6) = \sin x \cos \pi/6 + \cos x \sin \pi/6$$
$$= ([\sqrt{3}]/2) \sin x + (1/2) \cos x \text{ since } \cos \pi/6 = \sqrt{3}/2$$
$$\text{and } \sin \pi/6 = 1/2$$
$$= 0.5(\sqrt{3} \sin x + \cos x)$$

(c) Using the tangent of a sum of two angles formula:

$$\tan(x + \pi/3) = [\tan x + \tan \pi/3]/[1 - \tan x \tan \pi/3]$$
$$= [\tan x + \sqrt{3}]/[1 - \sqrt{3} \tan x] \text{ since } \quad \tan \pi/3 = \sqrt{3}$$

(d) Using the cosine of a difference of two angles formula:

$$\cos 3(x - \pi/18) = \cos(3x - \pi/6)$$
$$= \cos 3x \cos \pi/6 + \sin 3x \sin \pi/6$$
$$= ([\sqrt{3}]/2)\cos 3x + (1/2) \sin 3x$$
$$= (0.5)(\sqrt{3} \cos 3x + \sin 3x)$$

11.5 Without using a calculator evaluate:
  (a) $\cos 75°$ (b) $\sin 15°$

Solution:
(a) Using the cosine of a sum of two angles formula:

$$\cos 75° = \cos(45° + 30°)$$
$$= \cos 45 \cos 30 - \sin 45 \sin 30$$
$$= (1/\sqrt{2})(\sqrt{3}/2) - (1/\sqrt{2})(1/2)$$
$$= (\sqrt{3} - 1)/2\sqrt{2}$$

(b) Using the sine of a difference of two angles formula:

$$\sin 15° = \sin (45 - 30)$$
$$= \sin 45 \cos 30 - \sin 30 \cos 45$$
$$= (1/\sqrt{2})(\sqrt{3}/2) - (1/2)(1/\sqrt{2})$$
$$= (\sqrt{3} - 1)/2\sqrt{2}$$
$$= \cos 75°$$

11.6 Show that:
(a) $\cos 75° - \cos 15° = -1/\sqrt{2}$
(b) $\sin 75° + \sin 15° = \sqrt{(3/2)}$

Solution:
(a) Using the formula for the difference of two cosines:

$$\cos 75° - \cos 15° = -2\sin[(75 - 15)/2]\sin[(75 + 15)/2]$$
$$= -2\sin 30 \sin 45$$
$$= -2(1/2)(1/\sqrt{2})$$
$$= -1/\sqrt{2}$$

(b) Using the formula for the sum of two sines:

$$\sin 75° + \sin 15° = 2\sin[(75 + 15)/2]\cos[(75 - 15)/2]$$
$$= 2\sin 45 \cos 30$$
$$= 2(1/\sqrt{2})(\sqrt{3}/2)$$
$$= \sqrt{(3/2)}$$

Exercises

11.4 Expand each of the following:
(a) $\cos(x + \pi/3)$ (b) $\sin(x - \pi/4)$
(c) $\tan(x - \pi/4)$ (d) $\tan 4(x - \pi/12)$

11.5 Without using a calculator evaluate:
(a) $\sin 75°$ (b) $\cos 15°$

11.6 Find the value of:
(a) $\cos 75° + \cos 15°$ in terms of $\cos 15°$
(b) $\tan 30° + \tan 15°$ in terms of $\tan 15°$

The fundamental trigonometric identity
Pythagoras' theorem can be applied to similar right-angled triangles by scaling down the hypotenuse to unity. That is, given the right-angled triangle with side lengths a, b and c, *Pythagoras'* theorem states that:

$$a^2 + b^2 = c^2$$

By dividing both sides of this equation by c^2 we obtain the similarity equation:

$$(a/c)^2 + (b/c)^2 = 1$$

that is:

$$\sin^2 x + \cos^2 x = 1$$

This identity is called the **fundamental trigonometric identity**.

Further identities
Following from the fundamental identity are three others:
Dividing

$$\sin^2 x + \cos^2 x = 1$$

by $\cos^2 x$ yields:

$$\tan^2 x + 1 = \sec^2 x$$

Dividing

$$\sin^2 x + \cos^2 x = 1$$

by $\sin^2 x$ yields:

$$1 + \cot^2 x = \csc^2 x$$

Worked Examples

11.7 Expand $\cos 3x$ in terms of $\cos x$.

Solution:
Using the formula for the cosine of a sum of angles:

$$
\begin{aligned}
\cos 3x &= \cos(2x + x) \\
&= \cos 2x \cos x - \sin 2x \sin x \\
&= (\cos^2 x - \sin^2 x)\cos x - 2\sin x \cos x \sin x \qquad \text{using the double angle formulae} \\
&= (\cos^2 x - (1 - \cos^2 x))\cos x - 2\sin^2 x \cos x \qquad \text{using the fundamental identity} \\
&= (2\cos^2 x - 1)\cos x - 2(1 - \cos^2 x)\cos x \\
&= 2\cos^3 x - \cos x - 2\cos x - \cos^3 x \\
&= \cos^3 x - 3\cos x
\end{aligned}
$$

Exercises

11.7 Expand $\sin 3x$ in terms of $\sin x$.

Unit 3   The sine and cosine rules

Try the following test:

1   Given triangle ABC where $B = 45°$, $a = 16$ and $b = 12$, find angle A using the sine rule:

2. Given triangle ABC where $a = 9$, $b = 8$ and $c = 7$, find angle A using the cosine rule:

3   Given triangle ABC with sides a, b and c opposite angles A, B and C respectively solve the triangle for:
(a) $A = 15°$, $C = 80°$, $b = 20$ (b) $a = 32$, $c = 16$, $B = 120°$
(c) $a = 4$, $b = 2$, $A = 3\pi/5$ r   (d) $a = 11$, $b = 12$, $c = 13$

Solving triangles
All the properties of a right-angled triangle can be deduced when the length of any one of the sides and the size of any one of the two acute angles are given. However, this is not the case when the triangle is not right-angled. To assist in the deduction of the properties of any two triangles the sine and cosine rules have been devised.

The sine rule
The area of a triangle is:

$$(1/2)(\text{base})(\text{height})$$

Given the triangle shown in the following Figure:

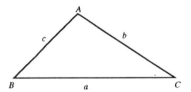

where sides a, b and c are respectively opposite angles A, B and C then the area of the triangle is given as:

$$(1/2)\ bc \sin A = (1/2)\ ac \sin B = (1/2)\ ab \sin C$$

If each of these expressions for the area is divided by $(1/2)abc$ then what results is called the sine rule:

$$\frac{\sin A}{a} = \frac{\sin B}{b} = \frac{\sin C}{c}$$

Worked Examples

11.8 Given triangle ABC where $B = 50°$, $a = 21$ and $b = 17$, find angle A using the sine rule:

Solution:
Using the sine rule we can say that:

$$\frac{\sin A}{21} = \frac{\sin 50}{17}$$

Hence:

$$\sin A = \frac{21 \times \sin 50}{17} \quad \text{so that } A = 71.14° \text{ to two decimal places.}$$

Exercises

11.8 Given triangle ABC where $C = 35°$, $a = 6$ and $b = 4$, find angle A using the sine rule:

The cosine rule
Given the triangle shown in the following Figure where h is the length of the perpendicular line from AB to C (one of the three **altitudes** of the triangle):

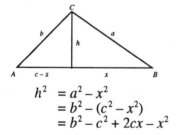

$$h^2 = a^2 - x^2$$
$$= b^2 - (c^2 - x^2)$$
$$= b^2 - c^2 + 2cx - x^2$$

Therefore:

$$a^2 - x^2 = b^2 - c^2 + 2cx - x^2$$

Adding x^2 to both sides yields:

$$a^2 = b^2 - c^2 + 2cx$$

That is:

$$b^2 = a^2 + c^2 - 2cx$$
$$= a^2 + c^2 - 2ac \cos B \text{ since } x = a \cos B$$

This is known as the cosine rule. Considering the other two altitudes of the triangle it is similarly shown that:

$$a^2 = b^2 + c^2 - 2bc \cos A$$

and

$$c^2 = a^2 + b^2 - 2ab \cos C$$

Worked Examples

11.9 Given triangle ABC where $a = 7$, $b = 5$ and $C = 85°$, find angle A using the cosine rule:

Using the cosine rule:

$$c^2 = a^2 + b^2 - 2ab \cos C$$
$$= 25 + 49 - 245 \cos 85$$
$$= 52.646...$$

Thus:

c = 7.26 to two decimal places.

Now we can use the sine rule:
$$\frac{\sin A}{7} = \frac{\sin 85}{7.26}$$
Giving $A = 73.85°$ to two decimal places.

Exercises

11.9 Given triangle ABC where $a = 7$, $b = 5$ and $c = 8.5$, find angle A using the cosine rule:

Application of the rules
Every triangle contains three sides and three angles. To find the values of all six items requires a prior knowledge of:

(a) two angles and one side length, or
(b) two side lengths and an angle opposite one of the given sides, or
(c) two side lengths and the included angle, or
(d) all three side lengths.

If the information given about a particular triangle is (a) or (b) then the sine rule can be applied to find the unknown side lengths and angles. If the information given is (c) or (d) then the cosine rule can be applied.

Worked Examples

11.10 Given triangle ABC with sides a, b and c opposite angles A, B and C respectively solve the triangle for:

(a) $A = 30°$, $B = 75°$, $c = 10$ (b) $a = 15$, $b = 25$, $C = 45°$

(c) $b = 17$, $c = 12$, $B = 2\pi/3$ r   (d) $a = 22$, $b = 20$, $c = 8$

Solution:

(a) $A = 30°$, $B = 75°$, $c = 10$

In triangle ABC we know two angles and one side length. Consequently we use the sine rule.

Since:

$$A + B + C = 10 + 75 + C$$
$$= 85 + C$$
$$= 180$$

we see that:

$$C = 95°$$

Hence, by the sine rule:

$$\frac{\sin C}{c} = \frac{\sin A}{a}$$

That is:

$$\frac{\sin 95}{10} = \frac{\sin 30}{a}$$

Therefore:

$$a = 10(\sin 30)/(\sin 95)$$
$$= 5.0 \qquad\qquad \text{to one decimal place}$$

Similarly:

$$\frac{\sin C}{c} = \frac{\sin B}{b}$$

That is:

$$\frac{\sin 95}{10} = \frac{\sin 75}{b}$$

Therefore:

$$b = 10(\sin 75)/(\sin 95)$$
$$= 9.7 \qquad \text{to one decimal place}$$

We have now solved the triangle *ABC*:

$A = 30°$
$B = 75°$
$C = 95°$
$a = 5.0$
$b = 9.7$
$c = 10$

(b) $a = 15, b = 25, C = 45°$

In triangle *ABC* we know two side lengths and the included angle. Consequently we use the cosine rule.

Since:

$$c^2 = a^2 + b^2 - 2ab \cos C$$

we see that:

$$c^2 = 15^2 + 25^2 - 2 \times 15 \times 25 \times \cos 45$$
$$= 319.7 \qquad \text{to one decimal place}$$

Therefore:

$$c = 17.9 \qquad \text{to one decimal place}$$

Similarly, since:

$$a^2 = b^2 + c^2 - 2bc \cos A$$

we see that:

$$15^2 = 25^2 + 17.9^2 - 2 \times 25 \times 17.9 \times \cos A$$

Consequently:

$$\cos A = (25^2 + 17.9^2 - 15^2)/(2 \times 25 \times 17.9)$$
$$= 0.8049...$$

Therefore:

$$A = 36.4° \qquad \text{to one decimal place}$$

Finally:

$$A + B + C = 36.4 + B + 45$$
$$= 81.4 + B$$
$$= 180$$

So that:

$$B = 98.6°$$

Thus completing our solution of triangle ABC.

(c) $b = 17$, $c = 12$, $B = 2\pi/3$ r

In triangle ABC we know two side lengths and one angle opposite one of the sides. Consequently we use the sine rule:

$$\frac{\sin B}{b} = \frac{\sin C}{c}$$

That is:

$$\frac{\sin 2\pi \sqrt{3}}{17} = \frac{\sin C}{12}$$

Therefore:

$$\sin C = 12(\sin 2\pi/3)/17$$
$$= 0.6113...$$

Therefore:

$$C = 0.6577... \text{ r } (or\ 37.7° \text{ to one decimal place})$$

From this we then find that (in radians):

$$A = \pi - 2\pi/3 - 0.6577$$
$$= 0.3894... \text{ r } (or\ 22.3° \text{ to one decimal place})$$

Finally, we can use either rule to find a. We continue with the sine rule:

$$\frac{\sin B}{b} = \frac{\sin A}{a}$$

That is:

$$a = 17(\sin 0.3894 \text{ r})/(\sin 2\pi/3 \text{ r})$$
$$= 7.4 \qquad\qquad \text{to one decimal place}$$

(d) $a = 22, b = 20, c = 8$

In triangle ABC we know the three side lengths and so we can use the cosine rule to find one of the angles. That is:

$$a^2 = b^2 + c^2 - 2bc \cos A$$

so that:

$$22^2 = 20^2 + 8^2 - 2 \times 20 \times 8 \times \cos A$$

so that:

$$\cos A = (20^2 + 8^2 - 22^2)/(2 \times 20 \times 8)$$
$$= -0.0625$$

Therefore:

$$A = 93.6° \qquad \text{to one decimal place}$$

Having found this angle we can now use either the sine or the cosine rule to find a second angle. That is:

$$\frac{\sin B}{b} = \frac{\sin A}{a}$$

so that:

$$\frac{\sin B}{20} = \frac{\sin 93.6}{22}$$

Hence $B = 65.1°$ (to one decimal place), giving $C = 21.3°$ (to one decimal place)

Exercises

11.10 Given triangle ABC with sides a, b and c opposite angles A, B and C respectively solve the triangle for:

(a) $B = 60°, C = 45°, a = 15$ (b) $b = 6, c = 12, A = 135°$
(c) $a = 35, c = 11, C = 3\pi/4$ r   (d) $a = 5, b = 7, c = 10$

Module 11 Further exercises

1   Find the sine, cosine and tangent of each of the acute angles in the half-equliteral triangle formed from an equilateral triangle of side length 4.

2   Use a calculator to find the sine, cosine and tangent of each of the following angles:
(a) 75°
(b) 1.05 r
(c) 8π/17 r
(d) 34.344°

3   Find the value of the secant, cosecant and cotangent of each of the following angles:
(a) 43°
(b) 0.95 r
(c) 13π/30 r
(d) 64.213°

4   Expand each of the following:
(a) $\sin(x + \pi/4)$
(b) $\cos(x - \pi/6)$
(c) $\tan(x - \pi/2)$
(d) $\sin 5(x - \pi/10)$

5   Without using a calculator evaluate:
(a) $\tan 75°$
(b) $\cos 5\pi/12$

6   Show that:
(a) $\sin 45° + \sin 15° = \cos 15°$
(b) $\cos 45° - \cos 15° = \sin 15°$

7   Expand $\sin 4x$ in terms of $\cos x$.

8   Given triangle ABC where $B = 33°$, $a = 10$ and $c = 15$, find angle A using the sine rule:

9   Given triangle ABC where $a = 22$, $b = 18$ and $c = 15$, find angle A using the cosine rule:

10 Given triangle ABC with sides a, b and c opposite angles A, B and C respectively solve the triangle for:
(a) $A = 75°$, $C = 65°$, $b = 13$
(b) $a = 11$, $c = 24$, $B = 135°$
(c) $a = 45$, $b = 29$, $A = 5\pi/6$ r
(d) $a = 9$, $b = 19$, $c = 29$

Appendix

The number represented by the numeral $\sqrt{2}$ – the square root of 2 – is not a rational number; it cannot be represented by a ratio of one integer to another. It is, accordingly, termed an **irrational** number.

The proof of this statement is quite straightforward and uses a method of proof known as **proof by contradiction**; we commence with an assumption and demonstrate that the consequence of accepting the truth of the assumption is that the assumption must be false.

1   *Assume* that the square root of 2 is a rational number and that it can be represented by the ratio of integers:

$$\sqrt{2} = \frac{m}{n}$$

where the ratio is in its lowest form – all common factors in the numerator and the denominator have been cancelled.

2   *Square* both sides of this equation:

$$2 = \frac{m^2}{n^2}$$

3   *Multiply* both sides of this equation by n^2:

$$2n^2 = m^2$$

4   *Conclude* that m^2 is an even integer because it can be divided by 2 to give n^2. Further, because m^2 is an even integer so is m; every even integer has an even square and every odd integer has an odd square.

5   *Write m* explicitly as an even integer in the form:

$$m = 2p$$

where p is an integer.

6   *Substitute* this form of m into the equation in 3 to give:

$$2n^2 = (2p)^2 = 4p^2$$

7   *Divide* both sides of this equation by 2:

$$n^2 = 2p^2$$

8   *Conclude* that n^2 and hence n are even integers and write n explicitly as an even integer in the form:

$$n = 2q$$

9   *Substitute* these two results into the original equation:

$$\sqrt{2} = \frac{m}{n} = \frac{2p}{2q} = \frac{p}{q}$$

10   *Conclude* that this result is inconsistent with the original assumption that the ratio *m/n* is in its lowest terms; we have produced a **contradiction**.

11   *Conclude* that the only way to resolve the contradiction is to deny the truth of the assumption – the assumption is therefore false and the square root of 2 is not a rational number.

Solutions

Solutions are provided for a selection of unit Test questions:

Module 1
Unit 1 1 (a) 339   (c) 91
       2 (a) 47

Unit 2 1 (a) 25751
       2 (a) 20   (c) 60
       3 (a) –25751
       4 (a) 91   (c)–91
       5 18; 1, 2, 3, 6, 9, 18. 45; 1, 3, 5, 9, 15, 45. HCF 9
       6 (a) 2, 3, 5, 7

Unit 3 1 (b), (c), (d)
       2 (a) 1/14
       3 (a) 9/14   (c) 3/11
       4 (a) 5/4   (c) –7/2
       5 (a) 10/7
       6 (a) 79/10   (c) 79/10
       7 (a) 37 1/2% (c) 22 2/9%
       8 (a) 1/5   (c) –3/20

Module 2
Unit 1 3 (a) 125   (c) 1
       4 2^n for n sets of copies made

Unit 2 1 (a) $2^9 = 512$ (c) $3^8 = 6561$
       2 (a) 1/32   (c) 81   (e) 7
       3 (a) 7772
       5 No

Module 3
Unit 1 2 (a) 0.4   (c) 4.25

       3 (a) $0.1\dot{7}1428\dot{5}$ (c) $-0.8\dot{3}$

Unit 2 1 (a) 2   (c) 3
       2 (a) 24.5, 24.45, 24.453   (c) 19.2, 19.19, 19.192
       3 (a) 24.5, 24.45, 24,453   (c) 19.2, 19.19, 19.192

4 (a) 6.32×10^{-6} (c) 3.00054×10^5
5 (a) 26102.8, 26103, 26000   (c) 1655.6, 1656, 2000
6 (a) 0.4, 0.3999...   (c) 0.875, 0.874999...

Unit 3 2 (e) 1.3018... (f) 0.0208...
        4 (a) 64/125   (c) 3/250

Module 4
Unit 1 1 (a) 5   (c) 27
        2 (a) 1001   (c) 11001001
        3 (a) 13   (c) 3010
        4 (a) 101   (c) 110001

Unit 2 1 (a) 10000111, 11111000, 11111001
          (c) 11101001, 10010110, 10010111

Unit 3 1 (a) 01010110   (c) 10010011
        2 (a) 01010110   (c) 01011111
        3 (a) 29   (c) –85
        4 (a) 11001100   (c) 10110010

Unit 4 1 (a) 11   (c) 15D
        2 (a) 76   (c) 3458
        3 (a) EE   (c) 2ED
        4 (a) 1101   (c) 1000110011100111

Module 5
Unit 1 1 960
        2 1
        3 1, 1, 2, 3, 5, 8, 13, 21, 34, 55, 89, ... Fibonacci sequence

Unit 2 1 (a) p, q, r; 1, –5, 3   (c) u, v, w; uv/w, u/w, v/w, uvw; 1/4, –3, 2, 1

Unit 3 1 (a) $4a + 8$ (c) $xy - 5xz$ (e) $gh - gk - gh - gk$
          (g) $6pq + 8pr - 8ps$
        2 (a) $x^5 y^{-2}$ (c) $a^2 c^{-2}$ (e) $g^{-3/4} h^{-3/2}$

Module 6
Unit 1 1 (a) $-9xz + yz - 3xy + 2y$ (c) $12lmn - 6mn$
          (e) $6s^2 t^4 + 6s^3 t - 7st^3 + 11st$
        2 (a) $-6m(2n + l)$ (c) $5b(c(a - 2) + (3a - 2))$

Unit 2 1 (a) $x(5z + 2y)$ (c) $k(2l(11j - 2) - 5j)$
          (e) $(s^3 + 3t^2 r)(r - t)$
        2 (a) $-3xy - 4x^2$ (c) $-18ac + 9bc$
          (e) $-32ac + 12bc - 21b^2 - 49ab$

Unit 3 1 (a) $35uw - 63xu - 15vw + 27xv$ (c) $15x^4 + 9x^3 + 7x^2 + 6x - 2$
   (e) $-5x^5 - 15x^4 + 22x^3 + 2x^2 - 20x + 16$
  2 (a) $(x - 2)(x - 5)$ (c) $(x + 8)(x - 8)$
   (e) $(x + 6)(x^2 - 6x + 36)$ (g) $(x - 2)(x + 5)$
   (j) $(11x - 7)(121x^2 + 77x + 49)$

Unit 4 1
$$
\begin{array}{ccccccccc}
 & & & & 1 & & & & \\
 & & & 1 & & -1 & & & \\
 & & 1 & & -2 & & 1 & & \\
 & 1 & & -3 & & 3 & & -1 & \\
1 & & -4 & & 6 & & -4 & & 1 \\
\end{array}
$$
  2 $a^6 - 6a^5b + 15a^4b^2 - 20a^3b^3 + 15a^2b^4 - 6ab^5 + b^6$
  3 (a) $a^5 - 5a^4b + 10a^3b^2 - 10a^2b^3 + 5ab^4 - b^5$
   (c) $a^4 - 2a^3b + 3a^2b^2/2 - ab^3/2 + b^4/16$
   (e) $256a^4 - 64a^3b + 6a^2b^2 - ab^3/4 + b^4/256$

Unit 5 1 (a) $2(b - 2a)/ab$ (c) $(y + 2x - 3)/xy$
   (e) $(9x^2 - 4y^2 + 36)/6xy$
  2 (a) $(1 - ab)/(b + 4)$ (c) $a - b$ (e) $2x + 3$
  3 (a) $2p$ (c) $a^2b^2/(a + b)$ (e) $(n + m)/(n - m)$

Module 7
Unit 1 1 (a) $y = 3x + 4$

x:	0	2	4	6	8
y:	4	10	16	22	28

   (c) $y = 2x^2 + 10x + 12$

x:	-4	-2	0	2	4
y:	4	0	12	40	84

   (e) $y = x^3 - 2x^2 - x + 2$

x:	-2	-1	0	1	2
y:	-12	0	2	0	0

Unit 2 1 $A (-6, 2)$, $B (4, 4)$, $C (6, -6)$, $D (-3, -5)$

Unit 3 1 (a)

x:	0.5	4	10/3	3/2
y:	-2.5	8	6	0.5

   (c)

u:	8/3	1/2	-7/3
v:	-10	-3.5	5

Unit 4 2 (a) $y = 3x - 7$ (c) $y = 6 - x/4$ (e) $y = 1 - 4x$

Module 8
Unit 1 1 (a) $x = (y + 4)/3$ (c) $n = [(m + 3)^{1/3} + 1]/2$
   (e) $q = [(p^3 + 9)^{2/3} - 4]^{1/2}$
  2 (a) $y = (3x - 9)/4$; $x = (9 + 4y)/3$
   (c) $u = 3(2 - (v/4)^2)^{1/2}$; $v = 4(2 - (u/3)^2)^{1/2}$
  3 (a) $-\sqrt{5} \le x, y \le \sqrt{5}$ (c) $f \ge \sqrt{2}$ or $f \le -\sqrt{2}$

Unit 2 1 (a) –1, –2, 6, 18   (c) 2, 0, 0, 0, –1
   (e) –1, 0, 0, 0, 0, 0, 0, 0, 1
  2 (a) $x = 6$ (c) $x = -3/2$ (e) $x = -3$
  3 (a) $x = -2, y = 3$ (c) $x = -4/3, y = 1/2$

Unit 3 1 (a) $x = 2$ or -3 (c) $x = -1/2$ or 1   (e) $x = \pm 2$
  2 (a) $x = (-3 \pm \sqrt{13})/2$ (c) $x = (5 \pm \sqrt{69})/2$ (e) $x = 1$ or $1/2$

Unit 4 1 (a) $x = -1 \pm \sqrt{2}$ (c) $x = (-7 \pm \sqrt{77})/2$ (e) $x = (1 \pm \sqrt{5})/2$
  2 (a) one (c) none (e) two
  3 (a) $k = \pm 2\sqrt{3}$ (c) $k = -9/4$ (e) $k = 0$ or $k = 28/9$

Module 9
Unit 1 1 (a) $x + 2$ (c) $3x + 4$ (e) $8x^2 + 12x - 22$ remainder 43

Unit 2 1 (a) $(x + 5)(x - 4)$ (c) $(x + 1)(x - 1)(x^2 + 4)$
  (e) $(x - 2)(4x - 1)(2x - 1)(x + 1)$

Module 10
Unit 1 1 (a) 0.2617... r   (c) 2.5440... r
  2 (a) 15° (c) 58.327...°
  4 $a = c, b = d, p = r, s = q : c,b$ a, d r, s p, q supplementary pairs
  5 (a) $\pi/4$ r (c) $-2\pi/3$ r

Unit 2 2 $C = 30°; B = C = 75°$
  3 $\sqrt{3}/2$

Unit 3 1 $AC = 20$
  2 $2\sqrt{3}$
  4 $6\sqrt{3}$

Module 11
Unit 1 1 $\sin A = 1/2, \cos A = (\sqrt{3})/2, \tan A = 1/\sqrt{3}$
   $\sin B = (\sqrt{3})/2, \cos B = 1/2, \tan B = \sqrt{3}$
  2 (a) $\sin 35 = 0.5735..., \cos 35 = 0.8191..., \tan 35 = 0.7002...$
   (c) $\sin 4\pi/9 = 0.9848..., \cos 4\pi/9 = 0.1736..., \tan 4\pi/9 = 5.6712...$
  3 (a) $\sec 67 = 2.5593..., \operatorname{cosec} 67 = 1.0863..., \cot 67 = 0.4244...$
   (c) $\sec 3\pi/7 = 4.4939..., \operatorname{cosec} 3\pi/7 = 1.0257..., \cot 3\pi/7 = 0.2282...$

Unit 2 1 (a) $(\sin x + \sqrt{3} \cos x)/2$ (c) $(\sqrt{3} \tan x - 1)/(\sqrt{3} + \tan x)$
  2 (a) $(\sqrt{3} + 1)/2\sqrt{2}$
  4 $8\cos^4 x - 8\cos^2 x - 1$

Unit 3 1 $A = 1.231$ r to 3 decimal places
  2 $A = 1.281$ r to 3 decimal places
  3 (a) $B = 85°, b = 5.20, c = 19.77$ (to 2 decimal places)
   (c) $B = 28.39°, C = 43.61°, c = 2.90$ (to 2 decimal places)

Index